AF885624

ORIGINS

Evidence for a Creation and Designed Universe

Facts and Wonder:
A Secular Exploration—From Skepticism to a Rational Case for Creation

Ralph Windsor

This book is dedicated to my father,
Benjamin Windsor

Copyright © 2025 by Ralph Windsor
All rights reserved. No part of this book may be reproduced, distributed, or transmitted in any form or by any means, including photocopying, recording, or other electronic or mechanical methods, without the prior written permission of the author, except in the case of brief quotations embodied in critical reviews and certain other non-commercial uses permitted by copyright law.

ISBN: **978-1-0692844-0-2**
Second Edition: 12, 2025
Printed in Canada and the United States

For permissions or comments, contact:
ralphwindsor1@outlook.com

Cover design: Ralph Windsor and Caryl Keating

A Note to the Reader: What This Book Is Not

Because this book investigates the evidence for a creation event, a natural assumption might arise: Is this a religious text in disguise? Is this a theological argument designed to lead to a specific faith?

The answer is an absolute and definitive no. The author is not religious, and does not subscribe to any faith-based traditions. This book carries no religious agenda, defends no traditional theology, and appeals to no spiritual dogma. It is written from a perspective that focuses strictly on what is solid, verifiable, and grounded in science.

Across the chapters of this book, an array of distinct disciplines are examined —including the precise mathematical tuning of physical constants, the digital-like architecture of quantum reality, the complex information code of DNA, and the emergence of conscious awareness.

When this book speaks of a "creation event" or a "deliberate origin," it is treating intelligence as a rational, data-driven scientific hypothesis based on the cumulative weight of all these fields—not as a matter of faith.

You will find no theological pivots or hidden doctrines. The data stands on its own. The inquiry is strictly scientific, the analysis is philosophical, and the conclusion is left entirely to your own rational judgment.

CONTENTS

"potential"

Merriam-Webster Dictionary, definition:

Existing in possibility; capable of development into actuality. Something that can develop or become actual; a possibility.

"We do not observe nature as it actually exists but nature exposed to our methods of perception."

Albert Einstein

INTRODUCTION

For most of my life, I believed that the cosmos arose through blind natural forces alone. I dismissed the idea of a created universe because I closely tied it to religious mythology. However, after years of study, I realized that was a mistake. When looking only at the evidence, one can recognize that the universe bears the unmistakable marks of intentional design without believing in a supernatural deity. My shift in perspective was driven entirely by data. I remain completely non-religious, but the evidence has convinced me that a creation event occurred—a rational hypothesis built on structure, logic, and code.

In this book, I use the words Creation and Creator differently than you might expect. To be clear, my shift in perspective was not based on any religious faith; I remain a non-religious person, and this is not a religious text. In fact, I believe religious doctrine obstructs a clear understanding of the data. Throughout history, religious narratives have provided supernatural explanations for our origins. Today, however, the idea of intentional design need not rely on religious belief. This book will neither preach nor proselytize. I consider myself a scientific creationist—someone who, after an objective examination of the evidence to determine whether a creation event occurred, believes there was such an event.

I see substantial evidence supporting that conclusion, which has motivated the writing of these chapters. I explore the evidence that the universe was deliberately formed, based on reason and observable facts. The purpose, meaning, and methodology of this creation are currently unknown—and remain open to interpretation. However, the fact of creation, based on the evidence, should be beyond doubt to any open-minded reader.

In the final chapter, I offer my personal reflections on creation and explore the simulation hypothesis as a speculative framework rather than a definitive explanation. This perspective represents how I currently think about these questions, while acknowledging that we cannot be certain about what and how creation took place.

I examine what a simulation, if true, might suggest about existence, meaning, and humanity's place in the universe. We must ask: if our reality is an engineered construct, what does that imply about the "Programmer" or the intent behind the code?

My intended audience is those who do not believe in Creation. I understand that viewpoint well. For decades, I saw the universe as a product of chance. But the deeper I explored, the more I found myself questioning that view. This book presents the case that the universe shows signs of design—an argument built not on mere belief, but on logic and empirical discovery. While readers with religious faith may find parallels here, this is not a religious text. It invites you to set aside any religious assumptions and re-examine the origins of existence from a fresh perspective.

Atheist skepticism toward any concept of a Creation is understandable because it is so closely tied to religion. Religion has long offered stories of divine intervention—some inspiring, others contradictory or implausible. When religious explanations conflict with empirical evidence or serve as vehicles for fear, moral control, or exclusion, it is natural to question them—and, by extension, to reject the idea of Creation altogether. Yet dismissing religious doctrine does not require dismissing the idea of intentional design. The flaws of human religious institutions do not negate the possibility that the universe was created with intention. Indeed, one can be critical of religion and still open to the idea that the cosmos may bear marks of intelligence.

Religious traditions have helped shape culture and moral thought, but they do not own the question of origins. If the universe was designed, the clues should be embedded in the structure of reality itself—not hidden in ancient texts, but revealed through the progress of science. My own journey began with some childhood religious belief but quickly shifted to disbelief. Raised with religious ideas, I eventually rejected them. I saw no compelling reason to accept the existence of a divine being. However, the idea of an intentional origin—not in a mythical sense, but as a rational hypothesis—became increasingly compelling. I found it difficult to accept that the universe, with all its complexity and precision, was simply a cosmic accident, devoid of guidance or purpose.

Science has illuminated vast realities that were once the exclusive domain of religious faith. While our understanding of physics, biology, and the cosmos has expanded exponentially over the past century, these discoveries have not undermined the case for design. Instead, they point to a level of complexity, balance, and order that cannot be brushed aside as pure randomness. Religion merely provided humanity's earliest frameworks for existence, but science now gives us the tools to examine the underlying mechanisms. Paradoxically, the deeper science looks, the more it reveals the exact, improbable precision of the universe's architecture.

The elegance of life is often cited as evidence of evolution, and rightly so. Evolutionary processes are well-supported by mountains of evidence from many independent fields; biological evolution is a fact. We observe evolution even today, in real time, especially in microorganisms that rapidly adapt and evolve into new strains. This dynamic adaptability is a testament to evolution and life's resilience. Yet the very mechanisms that drive evolution—mutation, selection, adaptation—are themselves deeply intricate. They unfold within the architecture of DNA, a molecule whose elegant structure encodes the instructions for all known life. These processes operate with remarkable precision, guided by biological systems of staggering complexity. Can such refined, tightly coordinated phenomena truly be the product of chance alone?

Rather than refuting Creation, evolution illuminates it. The very existence of such a robust, responsive system hints at an underlying architecture—something built to enable adaptation and survival. A framework that seems to have been anticipated. While evolution explains how life changes, it does not explain how life began. DNA, with its dense, information-rich structure, resembles computer code—ordered, purposeful, and exacting. It is not easy to dismiss it as a mere accident of chemistry. The presence of such intricate instructions within every cell suggests something beyond random assembly.

Modern physics, too, has expanded our understanding of reality in astonishing ways. Quantum mechanics, relativity, and the structure of spacetime reveal a universe far more mysterious and interconnected than previously imagined. These discoveries often defy intuition—and suggest that reality is not mechanical alone, but deeply layered, even purposeful. The Big Bang provides strong support for a creation event. From a singularity came space, time, matter, and energy. But more compelling than the event itself is what followed: the fine-tuning of physical constants—gravity, electromagnetism, nuclear forces. These values had to fall within extremely narrow ranges for stars, planets, and life to emerge. Even minor variations would render the universe lifeless.

Such precision is not easily explained by chance. It suggests deliberation. Purpose. In the chapters ahead, we will examine the scientific evidence that supports the idea of a created universe. These discussions will rely on testable data, critical reasoning, and observable phenomena—not religious assertions or philosophical claims. A singularity with immense potential became a structured cosmos. But potential alone does not organize itself. Something had to spark the transformation from formless energy into coherent systems—from particles to galaxies, from atoms to organisms, from unconscious matter to minds capable of asking these questions.

As you explore these ideas, I invite you to consider that our universe may not be the product of blind chance, but of purposeful creation. Einstein once said that a person exploring the mysteries of nature is like a child entering a vast library. The child sees the books, senses the order, but does not understand the language in which they are written. Still, the child knows someone must have written them. Our understanding is limited, but what we can observe is profound. And what we see points to more than myth—it points to something worth examining seriously.

Before the event of a Creation lies the idea of potential. Before the Big Bang, there wasn't nothingness—there was possibility. The universe's emergence wasn't merely an expansion, but a transformation: from potential into structure, from potential into reality. What force initiated that shift? What principle or intelligence enabled it? Where did this potential come from? These questions are still unanswered.

By the time you reach the end of this book, my hope is that you'll find the case for Creation compelling. Perhaps, you'll begin to see the universe not as a cosmic accident. And with that may come the realization that we are part of something far greater and awe-inspiring—than we could ever imagine.

Enjoy.

"Why is there something rather than nothing? This question may be beyond our capacity to answer."

Thomas Nagel

CHAPTER 1:

A New Perspective on Skepticism and Creation

In this chapter, presented in three parts, I begin with a critical examination of how dissatisfaction with traditional religious dogma—combined with an overconfidence in narrowly interpreted scientific explanations—has led many people, myself included, to dismiss the idea of Creation far too readily.

The modern intellectual climate often frames scientific inquiry and Creation as mutually exclusive positions, creating a false divide: one must either embrace science and reject Creation, or accept Creation and abandon rational inquiry. I argue that this division is neither necessary nor justified. It reflects not a conflict between evidence and belief, but a failure to distinguish between scientific methodology and the philosophical conclusions often drawn from it.

This is not an argument against science. On the contrary, science is an extraordinary and indispensable tool for understanding the mechanisms of the universe. It excels at uncovering the *how* of existence, revealing the laws that govern phenomena ranging from subatomic particles to the large-scale structure of the cosmos. Yet science, by its very design, is limited in its ability to address the *why*—questions of ultimate origin, purpose, and meaning. These questions are not rendered invalid by science; they simply lie beyond its methodological scope.

Over the past century, powerful cultural and intellectual forces have reinforced the assumption that rational thought inevitably leads to the rejection of Creation. Influential atheist writers, philosophers, public intellectuals, and media narratives have convincingly dismantled religious institutions and theological dogma. However, these critiques are frequently mistaken for decisive refutations of Creation itself. In reality, the collapse of institutional religion does not logically entail the absence of an originating intelligence or purposeful foundation to the universe.

I once accepted this conflation without sufficient scrutiny. Having been firmly grounded in atheism and disbelief in any form of Creation, I assumed that intellectual honesty required such a position. Yet when I returned to the evidence with greater care and fewer preconceptions, I found that this conclusion was no longer sustainable. What emerged instead was the recognition that Creation is not merely a theological assertion, but something that appears woven into the structure of reality itself—visible to those willing to look without ideological constraint.

Through this lens, Creation is not an antiquated belief, nor a retreat from reason, but a conclusion that arises naturally from sustained inquiry. By setting aside false divisions and cultural reflexes, we open the possibility of seeing reality more clearly. In doing so, we move toward a richer understanding—one that unites scientific insight with philosophical depth, reason with wonder, and knowledge with the enduring human search for meaning.

A: Disillusioned by Doctrine: The Atheist Journey

B: Challenging Faith: What Atheism's Leading Figures Taught Us Part

C: Reconsidering Creation: What led me to reconsider my views

Part A – Disillusioned by Doctrine: The Atheist's Journey

Atheism—and, by extension, the rejection of any belief in Creation—often arises not from a careful study of science, but from disillusionment with organized religion. This was certainly true for me, and I suspect it is true for many readers. Disappointment with rigid dogma, glaring hypocrisy, or even personal—sometimes painful—experiences can all contribute to turning away from faith. For many of us, this step feels necessary: a way to reclaim independent thought, to begin thinking for ourselves as we mature.

As noted in my introduction, evidence shows that atheists overwhelmingly reject religious teachings. Roughly 82% of atheists raised within a faith tradition consciously leave it behind. Because the idea of Creation is so tightly bound to religious doctrine, rejecting faith often feels inseparable from rejecting Creation itself. Yet this connection is not absolute.

Is it possible to reject religion while remaining open to the idea of a Creation—one grounded not in dogma, but in evidence? For many atheists, disbelief and the denial of a Creator are so entwined that separating the two seems impossible. I contend that it is not. When we focus purely on the scientific evidence explored in Chapters 2 through 6, arguments against the possibility of Creation begin to crumble.

Reflecting on my own journey, I realize my early doubts about Creation were born not from careful study, but from observing the contradictions and shortcomings in religion itself. In my younger years, I was influenced by atheist writers, compelling lecturers, and media narratives that reinforced disbelief. Science was often presented in ways that seemed to render Creation unnecessary, even obsolete. Among these ideas, the theory of evolution stood out as particularly persuasive. I concluded that if life could arise and develop through natural processes, then the notion of a Creator was redundant. Evolution became the bedrock of my conviction that Creation was neither required nor relevant.

For forty years, I viewed atheism simply as the logical conclusion of a mechanical cosmos. In a universe governed by accident, there is no grand blueprint or higher intellect to factor into our equations; human intelligence stands as the highest observer. This perspective offered a straightforward, scientific framework that was free from religious assumptions. My focus was entirely on what could be measured and proven.

Over time, however, I have come to understand that evolution and purposeful Creation are not mutually exclusive. Evolution describes an extraordinary process through which life develops and diversifies, and it is entirely plausible that a Creator could have chosen evolution as the mechanism by which life emerges. Seen this way, the scientific and the sacred can coexist, each illuminating a different facet of the same remarkable reality.

What first led you to question the existence of Creation? Was it the influence of religious teachings—or their absence? Personal experiences? The voice of secular thinkers? Reflection is critical. Understanding your journey is essential for engaging with the ideas in this book. You might wonder how examining your rejection of religion connects to considering Creation. I argue that it is central. In the chapters ahead, I will show that the universe itself reveals evidence of design. But to see it clearly, you must separate the evidence from the religious frameworks you have rejected. Only by approaching the question with curiosity, rather than reaction, can you appreciate what the universe has to reveal.

Adopting an open mind is rarely easy, particularly regarding deeply held beliefs. Confirmation bias—the tendency to seek information that reinforces our views—makes challenging entrenched thinking difficult. Over time, many atheists reinforce disbelief in private or social dialogues, further solidifying their positions. Research suggests that the path to unbelief often follows a recognizable pattern. For those raised in religious settings, the journey is longer: it begins with doubt, continues with criticism of religious practices, and culminates in rejection of faith and community.

My journey began like many others. I was born into a religious home, attended Sunday school, and went to church. My parents were not zealots; my mother was a devout, but basic believer, while my father was a more contemplative man. He didn't subscribe to religious doctrines, except for some lifestyle and moral issues he knew were relevant, yet he lived with an intuitive sense that there was something beyond the sunset—a reality larger than the physical world we could see.

Until the age of twelve, I followed teachings without question. Then, a single moment changed everything. During a Sunday school lesson on life after death, I asked about the "unreached"—those in distant lands who had lived and died without ever hearing the bible message. The teacher's explanation was fragile and logically unsatisfying; it implied a system that was inherently unfair. That moment didn't just spark doubt; it ignited a "logical fracture" in my foundation. I remember the walk home, feeling a sudden, intellectual childhood moment when I realized, I did not believe what they were teaching anymore. I realized the adults in the room didn't have the answers. I resolved to learn more.

As I studied religion over the years, I became convinced that it was entirely human-made. This realization didn't come from a place of cynicism, but from a deep, scholarly appreciation for the texts themselves. I began to view the Bible, and indeed the sacred texts of many traditions, not as divine dictation, but as rich, wonderful, and profound literature. I saw them as the ultimate expression of the human heart—our ancient ancestors' best attempt to codify morality, community, and the search for meaning. In addition, I saw it as means for those in power to keep control.

I explored the origins of these beliefs with the same curiosity one might bring to the study of Shakespeare or Melville. I examined the perspectives of various organizations and traced the evolution of thought through history, and the more I learned, the more clearly, I saw these ideas as brilliant products of the human imagination. They were mirrors of the cultures that produced them, filled with wisdom and poetry, yet undeniably earthbound.

Within that framework, rejecting the concept of Creation felt like a necessary sign of intellectual maturity. I had concluded that if the "manuals" for the Creator were man-made, then the Creator must be a man-made fiction as well. In my mind, the two were inseparable; to abandon the religious worldview was, by definition, to abandon the possibility of an intentional beginning. It would take decades for me to realize that while the literature of religion is a human map, the territory of the universe itself might still have an Architect.

Looking back, I am grateful for the discovery. My journey was never driven by cynicism but by curiosity. It became clear to me that it was religion itself—not science—that had initially convinced me there was no Creation. My fascination with religious study was profound; I was drawn to philosophy, history, the laws, rules and rituals. By studying religion, I gained insight into the complexity of human belief systems and the ways they shaped societies. This broadened perspective allowed me to approach religion with curiosity rather than dismissal. I recognized that while religion could inspire compassion and community, it was also frequently used as a tool for control.

These experiences deepened my skepticism but also cultivated empathy for those harmed by manipulative leadership. By twenty, I had developed the ability to engage in meaningful debates with devoutly religious individuals. Many assumed I misunderstood their teachings, but in almost every case, I found their arguments surprisingly fragile, their answers more exercises in deflection than genuine engagement with fundamental questions.

Throughout my life, I maintained an insatiable appetite for knowledge. I read extensively and critically examined every idea. Over decades, this pursuit solidified my identity as a confident atheist, convinced that the universe was a cosmic accident—a beautiful roll of the dice. Yet my early rejection of Creation was shaped less by science than by observing religion's contradictions. Ironically, it was these flaws within religion that cemented my disbelief. I had been so busy pointing out the inconsistencies of faith that I had ignored the impossible precision of the science.

Whatever your journey, I encourage you to approach the evidence in Chapters 2 through 6 with an open mind. Suspend preconceptions and re-examine Creation not through the lens of past dogma or reaction, but through observation, reason, and evidence. The universe has a story to tell—one that deserves to be seen clearly.

Part B – Challenging Faith: What Atheism's Leading Figures Taught Us

Atheists often find inspiration in thinkers whose ideas have profoundly shaped modern discourse on reason, skepticism, and unbelief. Figures such as Richard Dawkins, Christopher Hitchens, Stephen Hawking, Sam Harris, Carl Sagan, Bertrand Russell, Ayn Rand, and Madalyn Murray O'Hair have not only influenced my understanding of religion and atheism but also helped define the broader intellectual pursuit of truth. For much of my life, their works were more than just books; they were mentors. They taught me how to question, how to demand evidence, and how to remain steadfast in the face of societal pressure to "just believe."

The writings of these individuals resonate deeply with those who value rational thought and intellectual independence. Their courage in standing against dogmatic authority is something I still admire today. Yet, upon closer examination of their collective arguments, a recurring and somewhat startling theme becomes apparent.

Ayn Rand, for example, famously declared: "I want to fight religion as the root of all human lying... I want to prove that religion breaks a character before it's formed, in childhood, by teaching a child lies before he knows what a lie is."

Rand's statement exposes a pattern I had not fully appreciated during my decades of skepticism: while prominent atheist thinkers launch vigorous, necessary critiques of religion, they seldom pursue an equally rigorous examination of the possibility of Creation itself. Her rejection of religion becomes, by extension, an automatic rejection of any intelligent cause behind existence. While her determination to confront dogma is evident, her arguments—like those of many others—focus primarily on the institutional and moral failings of religion rather than on the deeper metaphysical questions concerning existence itself.

After re-examining the arguments advanced by these prominent atheists—ironically, the very individuals whose ideas I once championed in debates—I was struck by how limited their case against the possibility of Creation actually appeared. Despite their diverse backgrounds and disciplines, their conclusions often echoed one another in a specific way. Nearly every argument against Creation ultimately turned out to be an argument against the inconsistencies and contradictions of organized religion, rather than against the concept of a purposeful beginning.

Atheist writers often target the dogmatic aspects of religion—original sin, eternal punishment, or the authority of scripture—arguing that religious morality is deeply flawed. They frequently cite examples of intolerance, violence, or injustice perpetrated in religion's name. These critiques are often valid; they expose the excesses of human institutions. However, they rarely engage directly with the central question of Creation. Demonstrating the moral failings of religious leaders or the severity of doctrines such as Hell does little to address whether an intelligent cause underlies the mathematical precision of the universe.

Similarly, atheists often reject biblical literalism, emphasizing contradictions, inaccuracies, and pre-scientific worldviews within sacred texts. As I have stated, I view these texts as wonderful literature, but literature of human origin. Disputing the Genesis narrative is straightforward for any modern person, but such criticism alone does not constitute a rational engagement with the evidence for or against a Designer. We must ask: if we set the human books aside, what does the "book of nature" actually say?

A common argument in atheist discussions is that religion relies on a "God of the gaps" approach—using God to explain whatever science has not yet figured out. From this perspective, appealing to a Creator is seen as giving up on real investigation. The argument is persuasive, but I eventually realized it often flips into what I now call a "Science of the Gaps" mindset.

This is the assumption that science already has, or eventually *must* have, naturalistic answers to every deep question—including how the universe began, how life first emerged, and what consciousness truly is. This "Science of the Gaps" attitude does not disprove the possibility of Creation; it simply postpones the issue. Saying, "We don't yet know how life began, but science will discover the answer someday," assumes that a natural explanation is guaranteed. In reality, that is a statement of faith—not in a deity, but in a specific philosophical outcome.

Take abiogenesis, for example—the idea that life arose from non-living matter. Believing that future science will definitely solve this mystery may underestimate just how challenging these questions truly are. The appearance of life, the fine-tuning of the universe's physical constants, and the existence of consciousness are not merely residual gaps; they represent foundational enigmas. The extraordinary precision of the conditions necessary for life—the cosmological constant, the strength of gravitational forces, and the intricate molecular architecture of living cells—points to a level of coherence that science has yet to fully explain. Yet many atheist thinkers sidestep these fundamental questions, focusing their critique instead on the "low-hanging fruit" of religious absurdity.

Christopher Hitchens, perhaps the most eloquent of the "New Atheists," exemplified the intellectual vigor that made atheism such a compelling force. In *God Is Not Great: How Religion Poisons Everything*, Hitchens articulated a powerful critique of organized religion, contending that it fosters hypocrisy and suppresses inquiry. His rejection of religion rested on core claims expressed with passion: he regarded religion as a human construct designed to soothe existential anxiety and impose social order.

While these criticisms are rhetorically powerful—and I agree fundamentally with his assessment of religious institutions—they largely target religion's human manifestations rather than the metaphysical question of a Creator's existence. Even Hitchens acknowledged the limits of human understanding regarding the origins of the universe, yet he often used this uncertainty not as an invitation to deeper inquiry, but as a pivot back to the critique of religious faith. This effectively conflates opposition to organized religion with the denial of Creation—a conflation that weakens the philosophical rigor of his argument.

Hitchens frequently invoked evolutionary theory as a foundation for his rejection of divine explanations. The evidence for evolution is indeed overwhelming, but to reject Creation entirely because evolutionary theory challenges literal readings of scripture is to commit a categorical error. The existence of evolutionary processes does not preclude the possibility of an underlying creative intelligence; rather, it describes one of its mechanisms.

This tendency to "refute the map instead of the territory" is common. Lawrence M. Krauss, director of the Origins Project, offers a naturalistic perspective that downplays the need for a Creator, famously stating: *"Forget Jesus. The stars died so you could be born."* While this framing celebrates the grandeur of the natural world, it sidesteps the question of whether those processes might themselves originate from intentionality. In emphasizing the *mechanisms* of nature, Krauss avoids addressing the *origin* of those mechanisms.

A particularly illustrative exchange of ideas occurred in 2009 during a debate between Christopher Hitchens and John Lennox, the Oxford mathematician and philosopher, titled *"Is God Great?"* Watching this debate was a pivotal moment for me because it laid bare the two different levels on which this argument can be fought.

Hitchens, true to his form, concentrated almost entirely on the moral and historical shortcomings of Christianity. He spoke of the "slavery" of divine supervision and the "wickedness" of atonement. He posed his famous rhetorical challenge: *If there are thousands of religions, all claiming exclusivity, how can any of them be true?* It is a devastating point against human institutions, and one I found myself nodding along to for decades. Hitchens was effectively arguing that because the "mapmakers" (religion) were often corrupt or confused, the "territory" (a Creator) must not exist.

However, Lennox, as a mathematician, moved the goalposts back to the laboratory. He didn't just defend a church; he defended the rationality of the universe.

Lennox argued that the very fact that the universe is "intelligible"—that we can use mathematics to describe the movement of stars and the behavior of atoms—is a profound mystery that atheism struggles to explain. He famously pointed out that if our minds are merely the product of unguided, random processes, why should we trust our own capacity for reason? If the brain is just a "computer" that evolved solely for survival rather than for truth, then the scientist's confidence in their own logic becomes a form of faith itself.

This was the point where I saw Hitchens begin to pivot. While Hitchens was brilliant at exposing the "crimes" of the religious, Lennox was exposing the "limitations" of the materialist. Lennox argued that the fine-tuning of the universe—the precise settings of gravity, electromagnetism, and the nuclear forces—was not a "gap" in our knowledge, but a feature of the design. He suggested that it takes more "faith" to believe that these dozens of physical constants just happened to fall into the exact one-in-a-trillion range required for life, than to believe they were set by an intelligence.

Lennox's approach was a revelation to me. He wasn't asking me to believe in a specific miracle or a particular dogma; he was asking me to look at the mathematical signature of the cosmos. He differentiated between "Primary Causes" (an Architect) and "Secondary Causes" (the mechanisms like Evolution or the Big Bang). By doing so, he showed that Hitchens' attack on the *mechanisms* did absolutely nothing to disprove the *Architect.*

This conflation—treating the flaws of religion as evidence against Creation—is the "blind spot" of modern atheism. Hitchens' rhetoric was effective because it resonated with our shared sense of justice and our frustration with human hypocrisy. But as Lennox demonstrated, you can be a staunch critic of religious institutions while still being a logical proponent of an intelligent source. One does not logically negate the other.

Ultimately, the debate revealed that Hitchens was fighting a war against religion, while Lennox was making a case for reality. By setting aside the sociological distractions of the "New Atheists," we can finally focus on the foundational questions that Lennox raised: Why is there something rather than nothing? And why is that "something" so highly structured and mathematically accessible?

Ultimately, the critiques of religion advanced by figures such as Hitchens, Dawkins, and Rand have enriched public debate. Their influence has prompted us all to seek more coherent foundations for belief and morality. Yet their arguments, however incisive, often fail to confront the central philosophical issue: the possibility of a purposeful origin to existence itself.

The question of Creation transcends the boundaries of both religion and science. It invites an inquiry not merely into what is observable, but into *why* anything exists at all. As we turn to that question, the task is neither to defend religious institutions nor to cling to inherited certainties, but to approach the mystery of being with intellectual rigor and openness.

Part C – Reconsidering Creation

What led me to reconsider my views on Creation? How did I evolve from decades of disbelief to my current views and belief in Creation?

This transformation began about 20 years ago. Surprisingly, the catalyst was not a profound philosophical treatise, a sudden "bang on the head," or a mystical experience. Instead, it was something as unassuming as a nature documentary. In hindsight, that moment sparked a journey of exploration and introspection, driven by four key factors that profoundly challenged my views.

Firstly, I came to recognize that religion—not the concept of or evidence for a Creation—bore the brunt of criticisms—including my own—in nearly every rejection of the possibility that we were created. For forty years, my rejection of a Creator was, in reality, a rejection of the *people* who claimed to represent it. I had been committing a fundamental "category error." In my mind, the Architect and the tenants of the building were one and the same. If the tenants were hypocritical, cruel, or irrational, I concluded that the Architect must be a fiction.

I began to see an underlying distinction that even the most brilliant atheists, like Hitchens, seemed to overlook: the flaws of religion are not synonymous with an absence of evidence for a Creation. Human institutions are prone to corruption, power-seeking, and logical inconsistencies, but those are human traits, not cosmic ones. Dismissing the possibility of a Creator based on humanity's shortcomings struck me as a flawed approach. I realized that if I found a high-precision instrument in the desert, I wouldn't dismiss the idea that it was designed just because the person who showed it to me was untrustworthy. The "Machine" exists independently of the "Preacher." This epiphany planted a seed of curiosity, urging me to explore the idea of a Creation independently of any anti-religious biases.

Secondly, an introduction to quantum mechanics marked a profound turning point in my life. What began as a casual curiosity was sparked by my introduction to the double-slit experiment—a deceptively simple test that produced results so bizarre they seemed almost magical. Intrigued, I dove into the fascinating world of quantum physics, a field that would radically challenge my understanding of reality.

The double-slit experiment, at its core, examines how particles such as electrons or photons behave when passed through two narrow slits. When no attempt is made to determine which slit a particle travels through, the particles display wave-like behavior, producing an interference pattern on a detection screen. This pattern is a clear signature of waves rather than an expected band of individual particles.

When a measuring device is introduced to determine the particle's path, the behavior changes. The interference pattern disappears, and the particles instead arrive at the screen as localized impacts, consistent with particle-like behavior. This challenges every classical intuition I held. It's not that the particles "know" they are being watched; rather, it is about whether the particle interacts in a way that leaves a physical record. This suggests that at the most fundamental level, reality is sensitive to interaction and information.

The sheer elegance and interconnectedness of quantum phenomena defied the notion of randomness I once believed governed everything. Was it truly reasonable to ascribe such a deeply ordered yet profoundly enigmatic system to blind chance? The more I learned, the more it seemed that quantum mechanics hinted at something deeper—a layer of intentionality embedded in the very fabric of reality.

Thirdly, my perspective was profoundly altered by David Attenborough's *The Private Life of Plants*. This captivating documentary series, using groundbreaking time-lapse photography and accelerated video footage, revealed a hidden world of plant behavior I had never imagined. For someone who had viewed plants as little more than passive, stationary objects, this was a revelation. It was as if a veil had been lifted; the plants in the time lapse sequences took on mammal-like qualities in their relentless drive to survive, compete, and reproduce.

Watching this series, I was captivated by what can only be described as a quiet, organic intelligence. Through the magic of the time-lapse, I saw plants engaged in slow-motion battles for territory. I watched as vines used their tendrils like sensitive "fingers," reaching out into the air, circling and searching until they made contact with a support, at which point they would instantly coil with the precision of a lasso. It was a strategic, goal-oriented movement that felt entirely purposeful.

Far from being static, these organisms were shown to be dynamic and resilient. I saw roots that didn't just grow downward, but stretched with uncanny accuracy toward hidden water sources, as if they could "sense" the moisture from a distance. I watched blossoms time their blooming to coincide precisely with the arrival of specific pollinators, often using complex chemical lures or physical traps. This was not the behavior of a "biological accident"; it was a refined, built-in wisdom, honed and encoded into the very biology of the species. The more I learned, the more it seemed that intelligence was not confined to brains—it was embedded in life itself.

If even a seemingly simple organism like a plant demonstrates such intricate mechanisms and finely tuned adaptations, what does that imply about the larger systems of nature? Was the apparent purposefulness in these processes a mere illusion, or was it evidence of a deeper intentionality within nature? As a skeptic, I had always believed in a universe of blind chance, but the "intelligence" of the green world suggested that the "code" of life was written with a specific intent in mind.

Finally, a personal family circumstance compelled me to confront the fragility and wonder of life and death. These moments of profound introspection have a way of stripping away intellectual pretenses. I found myself standing at a crossroads where the cold, indifferent answers of my atheism—that we are merely "biological accidents" in a random cosmos—could no longer satisfy the deep questions of meaning and purpose that were now knocking at my door.

This emotional and intellectual shift propelled me into a much deeper study of the very science I had always taken for granted. For much of my life, I was a "passive believer" in the scientific consensus; I simply assumed that the mystery of life's origins had been solved decades ago. I believed, as many do, that science had it all figured out—that there was a tidy, well-proven explanation for how a chaotic, prehistoric soup transformed into a living, breathing cell.

Yet, as I dug deeper into the actual data, I was astonished to find that science hasn't got a clue about how life could have evolved on its own. The transition from non-living matter to the first self-replicating molecules remains one of the greatest enigmas in human history. I had expected to find a clear path of evolutionary "stepping stones," but instead, I found a massive, unexplained leap. I began to look at DNA not just as a biological molecule, but as the fundamental "software" of existence.

The precision of the information encoded in DNA—the building blocks of life—defies an accidental explanation. We are talking about thousands of pages of complex, functional instructions that must be exactly right for life to even begin. In any other context—whether it is a computer program or a piece of literature—we immediately recognize that such high-level, specified information requires an intelligent input. Yet, in biology, I had been trained to believe that "blind chance" was a sufficient explanation.

I had reached a point where it required more "faith" to believe that this digital code wrote itself than to believe it was the work of a Creator. This realization was the final weight that tipped the scales. My atheism could be traced back to the age of twelve when a simplistic and dismissive Sunday school lesson left me disillusioned. I realized how unfair it was to let the shortcomings of human institutions, or the vile concept of Hell—a tool of torture used to frighten children—dictate my stance on such a profound question.

Separating these human failures from the concept of a Creation allowed me to approach existence with fresh eyes. While I view the Bible and other religious texts as profound literature—the ultimate expression of the human heart—they are products of human imagination. However, the universe itself appears to be a masterpiece of intricate design.

This journey led me to the logical conclusion that there was, in fact, a creation—and, by extension, a Creator. This marked my transition to becoming a Scientific Creationist. This title reflects my commitment to evidence-based inquiry, untethered from human religious doctrines. I am not looking for a god in a religious book; I am looking for the Architect in the evidence.

"The Big Bang is as close as we can get to a Creation event in science. It tells us that the universe had a beginning and has been evolving ever since."

Lawrence Krauss

CHAPTER 2

The Big Bang and the Seed of Creation

What is Cosmic Potential and Where Did It Come From?

Let us now examine the evidence pointing toward a creation event—evidence that, if taken seriously, strongly suggests the involvement of an intelligent source. Consider a question that seems simple at first glance yet carries extraordinary weight: Did the universe emerge by sheer chance, or was it purposefully created? The question is deceptively straightforward, but its implications challenge our deepest assumptions about reality and our place within it.

This single inquiry draws us into some of the deepest mysteries of cosmology, physics, and metaphysics. When we attempt to explore it scientifically—without appealing to religious dogma or pre-established belief systems—we find that a surprising pattern emerges: the evidence increasingly points not toward randomness, but toward an underlying order, structure, and intentionality woven into the fabric of the cosmos.

Modern science has delivered an astonishing wealth of insights into how the universe began, and many of these insights are not easily reconciled with the notion of a purely accidental origin. At the center of this story is the Big Bang.

According to decades of astronomical observations and cosmological modeling, the universe began approximately 13.8 billion years ago in an unimaginably dense and hot state commonly referred to as a singularity. This singularity is not merely a convenient theoretical construct; it arises as the natural conclusion of tracing the universe's continual expansion backward in time. As we extrapolate this expansion, all matter, energy, space, and even the fabric of spacetime itself converge toward a point where our current mathematical models and known laws of physics effectively break down.

The concept of the singularity is profoundly significant because it represents the boundary at which classical physics fails and a deeper explanatory framework becomes necessary. Within this infinitesimal state, densities and temperatures reach extremes that make our current theories—General Relativity and Quantum Mechanics—mutually incompatible. In other words, the singularity marks not just the beginning of the universe, but also the limit of human scientific understanding. Its existence suggests that whatever initiated the universe lies beyond the mechanisms we can currently describe, raising fundamental questions about causation, origins, and the possibility of an intentional creative act. From this initial point, space, time, energy, matter, and the fundamental laws themselves unfolded.

This expansion was first discovered by Edwin Hubble in the 1920s. By examining the light from distant galaxies, Hubble found that the farther away a galaxy is, the faster it appears to be moving from us. This observation revealed that space itself is stretching—a discovery that overturned centuries of assumption that the universe was static and unchanging.

If the cosmos is expanding now, then at some time in the distant past, everything must have been compressed together. This reasoning forms one of the foundations of the Big Bang model. Modern telescopes, including space-based observatories, have confirmed Hubble's findings with extraordinary precision, mapping out the expansion of space across incomprehensible distances.

Additional confirmation comes from the Cosmic Microwave Background (CMB), the faint afterglow of the early universe. In 1964, scientists detected this radiation accidentally. What they discovered was the thermal signature of the universe when it was only about 380,000 years old—after it cooled enough for atoms to form and light to travel freely.

This background radiation blankets the universe uniformly, yet contains subtle temperature variations. These minuscule fluctuations reflect the first density variations that later grew into galaxies, stars, and planets. The CMB not only matches theoretical predictions but does so with such remarkable fidelity that it has become one of the most powerful pieces of evidence for the Big Bang. Its existence, structure, and characteristics point toward an ordered beginning rather than a chaotic or random one.

The theory of Big Bang nucleosynthesis deepens this picture. In the first minutes after the universe began, temperatures were high enough for nuclear reactions to occur, creating the lightest elements: hydrogen, helium, and small amounts of lithium. The predicted proportions of these elements—based on known physical laws—match observations with extraordinary precision. This remarkable alignment between prediction and observation strongly suggests that the early universe operated according to stable, finely balanced laws.

The universe's large-scale structure provides yet another layer of evidence for underlying order. Galaxies are not scattered haphazardly throughout space; instead, they form an intricate web-like structure composed of filaments, clusters, and vast cosmic voids. This pattern emerged from initial quantum fluctuations in the universe's earliest moments. The existence of such a structure indicates that the universe did not evolve arbitrarily. From the earliest fraction of a second, it possessed precisely the right initial conditions to give rise to the ordered complexity we now see.

Then there is the discovery that the universe's expansion is accelerating, driven by a mysterious phenomenon known as dark energy. This acceleration was revealed in the late 1990s through the study of distant supernovae. The fact that this acceleration is neither too strong nor too weak is striking. If it had been slightly different, galaxies would never have formed, stars would not ignite, and life could not possibly arise.

The delicate balance of cosmic expansion, gravity, and dark energy suggests a finely tuned system operating within narrow parameters—so much so that, as astrophysicist George Ellis notes, the realization of the complexity involved "makes it very difficult not to use the word 'miraculous.'" This observation underscores just how extraordinary the harmony of the universe truly is, hinting at a cosmos uniquely poised to allow life to emerge.

The strength of gravity, the mass of elementary particles, the charge of the electron, the ratios of the forces, and the value of the cosmological constant all fall within extraordinarily precise ranges. Even slight deviations in any one of these constants would produce a universe that collapses instantly or expands too rapidly for structure to emerge. The improbability of these constants aligning by chance is so great that many scientists compare it to throwing darts blindfolded across the universe and hitting the same precise point every time.

A helpful analogy is the DNA molecule. DNA contains encoded instructions capable of generating complex life through the arrangement of its nucleotides. In a similar way, the constants of physics seem to form a kind of cosmic code, governing the formation of atoms, stars, galaxies, and eventually living organisms.

Quantum mechanics adds even more layers to this mystery. At the subatomic scale, particles exist in probabilistic states rather than in fixed positions. Quantum mechanics suggests that "observation"—meaning a physical interaction or measurement that creates a record—helps determine a particle's state. The rules of quantum behavior are astonishingly precise. These rules allow quantum fluctuations to seed cosmic structure, allow atoms to form stable configurations, and allow the universe to produce complex molecules, including those necessary for life. The quantum world is not chaotic; it operates according to subtle, elegant, but deeply unintuitive rules. This precision suggests underlying order far more than randomness.

In response to the problem of fine-tuning, some physicists have proposed the multiverse hypothesis: the idea that our universe is only one of countless universes, each with different constants. While this hypothesis is intriguing, it raises further questions: what mechanism generates these universes? What laws govern the multiverse itself? And how did those laws come to be? Invoking an infinite number of universes does not eliminate the need for an explanation of the order and potential within the overarching system.

This brings us back to the singularity. The singularity was not simply raw, featureless energy; it was a concentrated point that contained the laws, constants, and potential that would shape the cosmos. From this point, the universe unfolded with extraordinary precision. The existence of such potential implies a cause capable of producing it. Nothing in our experience suggests that order, laws, or potential simply arise spontaneously from absolute nothingness.

Philosophically, the question becomes one of ultimate causation. Science is powerful in explaining how the universe evolves from given starting conditions, but it does not—and cannot—explain why those starting conditions exist. If the universe displays order, coherence, and intelligibility, it is reasonable to consider the possibility that these characteristics reflect an intelligent origin. This conclusion does not affirm any religious narrative; it simply acknowledges the implications of the evidence.

Some modern thinkers, inspired by advances in computing, have proposed the simulation hypothesis—the idea that our universe might be a simulation created by a higher-level intelligence. While unproven at this time, it offers a striking perspective on intelligent design, showing that the universe's remarkable precision need not point to traditional theology.

Personally, I am inclined toward this view, not only because of the delicate balance of cosmic forces and the emergence of life, but also because of the astonishing facts revealed by quantum reality—the strange, counterintuitive behavior of matter and energy at the smallest scales. Taken together, these phenomena suggest a universe structured with extraordinary care. The simulation hypothesis invites us to entertain the possibility that intelligence underlies the cosmos in ways both profound and mysterious. It is a recognition that the universe behaves in ways suggestive of information processing, mathematical structure, and intentional parameters. Quantum mechanics, with its discrete energy states and "digital-like" behavior, adds to the intrigue.

Even if, for now, we set aside the simulation idea, the universe itself displays unmistakable order. Physical laws are mathematically precise and stable across time and space. Stellar evolution follows predictable patterns. Chemical reactions behave consistently. Life arises under conditions so specific that the coincidence of these conditions occurring without guidance appears extraordinarily unlikely. Einstein and many other great scientists marveled at this mathematical coherence, seeing in it something akin to a deeper rationality.

Biology further reinforces this theme. DNA is a code, not merely a chemical chain. Evolution explains adaptation and diversification but does not address why the universe has the exact conditions needed for chemistry, complexity, and life to exist. For evolution to operate at all, the universe must already be finely tuned.

Finally, the emergence of consciousness is perhaps the most striking evidence of all. In a universe that could have been sterile, lifeless, or chaotic, conscious beings arise—capable of curiosity, reasoning, creativity, morality, and wonder. When we consider all these elements together—the Big Bang, the CMB, nucleosynthesis, cosmic structure, dark energy, fine-tuning, quantum mechanics, biological information, and consciousness—a coherent picture emerges.

The evidence does not point to randomness. It points toward intention. It points toward order. It points toward a universe whose origin and structure appear aligned with deliberate creation. Rejecting this conclusion requires assuming that an extraordinary series of improbable events occurred without cause, direction, or purpose. While science cannot yet identify the ultimate source, the inference from the evidence is strong: the universe's origin and design are far more consistent with a deliberate cause than with blind chance.

Even if the nature of this cause remains beyond our full comprehension, the universe invites us to recognize its remarkable precision. From the singularity's potential to the emergence of galaxies, stars, and conscious beings, the cosmos reflects an underlying structure that is mathematically elegant and deeply meaningful. In contemplating these discoveries, we are reminded that the universe is not random matter drifting through space, but a coherent, intelligible system.

The universe challenges us to consider that behind all existence lies a deliberate origin—scientifically grounded, rationally sound, and profoundly compelling. All that we see, and all that we are, trace their lineage back to what emerged at that singular beginning: the Big Bang event.

"Mathematics is not just a tool for understanding the universe; it is a very deep part of the structure of reality itself."

Roger Penrose

"The elegance of mathematics is not only in its power to solve problems but also in its ability to describe the underlying structure of the universe. To me, that is evidence of a Creator who designed it all."

John Lennox

CHAPTER 3

The Discovered Mathematical Perfection of the Universe

From the moment we open our eyes to the world, patterns surround us—patterns that speak a silent language of order and beauty. Whether it is the symmetry of a butterfly's wings, the rhythmic waves of the ocean, or the spiral of a seashell, these patterns hint at a deeper structure underlying reality. Mathematics, often thought of as a purely abstract pursuit of numbers and formulas, is in fact the key to unlocking this hidden order. It is not confined to textbooks or classrooms; it is alive in the very fabric of creation, waiting to be noticed by those who are curious enough to look.

What strikes me most about mathematics is how it reveals truths that are at once simple and profound. These truths are not hidden—they are there whether in nature, in the cosmos, or in the intricate architecture of life itself. Mathematics provides a lens through which we can understand the universe, offering insights that go beyond human imagination. It is both universal and timeless, a language written into the cosmos long before humans existed to interpret it.

Among the most captivating mathematical patterns found in nature is the Fibonacci sequence. This is a series of numbers where each number is the sum of the two that come before it: 1, 1, 2, 3, 5, 8, 13, 21, 34, and so on. At first glance, it seems like a simple curiosity, a playful game of numbers. But look closer, and you begin to see it everywhere.

In the world of botany, this is known as Phyllotaxis—the study of the arrangement of leaves, seeds, and petals. If you look at a sunflower, you will notice that the seeds are arranged in two sets of spirals, one winding clockwise and the other counter-clockwise. If you count these spirals, you will almost always find two consecutive Fibonacci numbers—perhaps 34 and 55, or 55 and 89.

This is not a mere "aesthetic choice" by the plant. It is a masterpiece of engineering. By following this mathematical arrangement, the sunflower ensures that its seeds are packed together with the highest possible efficiency, leaving absolutely no wasted space. Furthermore, the leaves of many plants grow at an angle of precisely 137.5 degrees relative to the one below it. This is known as the Golden Angle, derived from the Golden Ratio. Why this specific number? Because it is the only angle that ensures no leaf perfectly overlaps the one below it, allowing every single leaf maximum access to sunlight and rain. This level of "biological foresight" suggests that the math preceded the plant.

Closely related to the Fibonacci sequence is the Golden Ratio, roughly equal to 1.618 (represented by the Greek letter Phi). This ratio appears when you divide one Fibonacci number by the one before it. It creates a proportion that the human eye perceives as inherently "right" or "beautiful." Ancient Greeks designed the Parthenon with this ratio; Renaissance masters like Leonardo da Vinci applied it to the proportions of the human body in the *Vitruvian Man.*

Yet, the Golden Ratio was not invented by artists; it was discovered by them. It is present in the spirals of a nautilus shell, the curve of a breaking wave, and even the proportions of the human face. The repeated appearance of these patterns across both human design and nature raises a striking question: Why should these patterns occur so consistently? Are they mere coincidences, or do they hint at a deeper order embedded in creation? John Lennox, a mathematician and philosopher, described these patterns as “brushstrokes on the canvas of the universe,” reflecting a rational and purposeful design.

As we step deeper into the structure of reality, we encounter fractals—endlessly repeating patterns that look the same no matter how closely you zoom in. In nature, they are everywhere: the jagged edges of coastlines, the branching of trees, the veins of leaves, the spiral of a nautilus shell, and even the delicate structure of snowflakes.

Fractals are not merely beautiful—they are a mathematical solution to a physical problem. Consider the human lung. To keep a human being alive, the lungs must absorb a massive amount of oxygen. To do this, they need a vast surface area. If our lungs were simple balloons, they would have to be the size of a room to function. Instead, they use a fractal branching pattern. One large airway branches into two, which branch into four, then eight, and so on, through 23 levels of branching. This allows a surface area roughly the size of a tennis court to be packed into a small, portable chest cavity.

The same principle applies to our circulatory system. Thousands of miles of blood vessels are packed into our bodies using fractal geometry to ensure every single cell receives nutrients. Even the way a tree branches follows this fractal logic to maximize the "catchment area" for sunlight while maintaining structural integrity. Everywhere we look, fractal patterns optimize function while creating awe-inspiring beauty. This suggests that the universe was designed by an "Efficient Engineer" who used mathematics to solve the problems of space and energy.

Nowhere is the precision of mathematics more striking than in DNA, the molecule of life. We often think of DNA as a biological substance, but it is more accurate to think of it as a digital code. While our computers use a binary system (0 and 1),

DNA uses a quaternary system of four chemical bases: adenine (A), thymine (T), cytosine (C), and guanine (G).

The amount of information packed into each tiny cell is staggering. Over two meters of DNA are folded into a microscopic nucleus without ever becoming tangled or knotted. How is this possible? It turns out that DNA uses a specific mathematical folding pattern known as a Hilbert Curve. This fractal-like folding allows the code to be incredibly compact while remaining perfectly accessible for the "reading" machinery of the cell.

Furthermore, DNA contains error-correcting mechanisms that mirror the "checksums" used in modern computer science. When the code is copied, enzymes "proofread" the sequence to ensure accuracy. Francis Collins, one of the leaders of the Human Genome Project, described DNA as a “language” written in code. This language is so precise and sophisticated that it mirrors systems human engineers spend years attempting to replicate. The consistency and reliability of DNA’s structure and function point to an underlying intelligence woven into the design of life, far beyond the reach of random chance.

If the patterns of life and nature inspire awe, the motion of the heavens has fascinated humans for millennia. The study of planetary orbits, tides, and gravity all reveal a universe governed by precise mathematical laws. No figure embodies this revelation more than Sir Isaac Newton, whose work bridged human curiosity and cosmic order.

Newton formulated the laws of motion and universal gravitation, showing that the same principles governing an apple falling from a tree also dictate the movement of planets. He described the universe as operating with clockwork precision, each celestial body following predictable paths determined by exact mathematical relationships.

What is remarkable is that Newton saw this order as evidence of a Creation. He wrote that the harmony and predictability of the cosmos pointed to "the most beautiful system of the sun, planets, and comets" as the work of "an intelligent and powerful Being." To Newton, mathematics was not just a human invention but a window into Creation. He believed that the universe was so finely balanced that a Creator must constantly "sustain" the laws of physics to prevent the system from falling into chaos.

Newton's insight into gravity also revealed hidden harmonies in planetary motion. Planets orbit the sun in ellipses described by precise ratios and speeds. Moons and satellites, like those of Jupiter, display orbital resonances—locked in numerical relationships (like 1:2:4) that prevent gravitational interference. The precision is staggering: even tiny deviations in gravitational constants would destabilize planetary orbits and make life on

Earth impossible. Newton recognized that such perfect calibration suggested a universe designed with intention.

What deepens this mystery is that mathematics does not merely describe what we can already see; it routinely predicts what we have yet to discover. Time and again, physicists have sat at desks working purely with abstract geometry and equations, uncovering mathematical truths that seemed to have no physical reality—only for experimentalists to find them decades later out in the cosmos. Black holes, gravitational waves, and subatomic particles were all mathematical certainties on paper long before they were observed by human technology. If mathematics were simply a human invention, a tool we manufactured to organize our thoughts, it could never possess this extraordinary predictive power. It acts not as a mirror reflecting our own minds, but as a map tracking an objective reality that was laid down before us.

This brings us to a profound philosophical question: Why does math work at all? In 1960, the physicist Eugene Wigner wrote a famous paper titled *"The Unreasonable Effectiveness of Mathematics in the Natural Sciences."* He argued that the fact that mathematical formulas—which are products of human thought—can perfectly predict the behavior of physical objects (like planets or subatomic particles) is a "miracle" that we neither understand nor deserve.

If the universe were a random accident, there is no reason why it should be "comprehensible." Our brains should have evolved only to find food and avoid predators; they should not be capable of understanding the curvature of spacetime or the internal mechanics of a star. The fact that the universe is written in a language we can read implies that we were meant to read it.

Mathematics is the bridge between the human mind and the cosmos. It shows us that the universe is not chaotic, but orderly; that beneath apparent complexity lies elegant simplicity. As Einstein famously remarked, “The most incomprehensible thing about the universe is that it is comprehensible.”

As we step back from the intricate details—from the spirals of sunflowers to the orbits of planets—we see a remarkable picture. Mathematics is not an invention; it is a discovery. It is the language of creation, written into the cosmos long before we arrived.

From the microscopic world of DNA to the vast expanse of galaxies, the universe displays harmony, balance, and astonishing precision. The Fibonacci sequence, the Golden Ratio, fractals, symmetry in physics, and fine-tuned constants all point to a cosmos that is intelligible and coherent. These patterns do not merely facilitate life—they allow us to understand life and the universe itself, suggesting a reality designed with intention and intelligence.

In recognizing the pervasive presence of mathematical order, we are invited to glimpse the hand of a Creator. This Creator, having woven mathematics into the very fabric of reality, provides a universe that is not only habitable but intelligible, accessible, and breathtakingly beautiful. The universe challenges us to consider that behind all existence lies a deliberate origin—Mathematics proves that the cosmos is not a silent accident, but a grand symphony of logic, proportion, and purposeful design.

"DNA is like a computer program but far, far more advanced than any software ever created."

Bill Gates

"The nitrogen in our DNA, the calcium in our teeth, the iron in our blood, the carbon in our apple pies were made in collapsing stars. We are made of star-stuff. "

Carl Sagan

CHAPTER 4

Evolution and the incredible complexity of DNA

Looking back on my many years as an atheist, there was one consideration that never wavered: the issue of evolution. Evolution was not merely an academic topic for me; it was a cornerstone of my atheistic beliefs. Like a foundational pillar, it supported my views, lending weight to my understanding of life's origins and the natural processes shaping existence. In that, if we evolved, which we did, then a Creation was unnecessary.

For many readers, I suspect this resonates deeply—evolution feels unassailable in its certainty, a steadfast framework for understanding the natural world. Now it may surprise you that my understanding and belief in the fundamentals of evolution still has not wavered. As I mentioned in the introduction to this book, one might assume that my current belief in a Creation—or a Creator—would inevitably challenge or contradict my long-held evolutionary convictions.

It seems logical to think that embracing the idea of a Creation might require rejecting evolution, aligning instead with views often associated with religious doctrines. Yet, as I emphasized earlier, any religious interpretations you draw from this work will be entirely your own. My goal here is not to present a theological argument but rather to explore the intersections of belief in a Creation and scientific evidence.

Still, the question persists: how can I now subscribe to both a Creation and a Creator while maintaining my certainty in evolution? On the surface, these concepts might seem incompatible, even mutually exclusive. For decades, the cultural and intellectual narrative has often presented evolution and Creation as two opposing forces in a zero-sum debate. But what if this contradiction is unnecessary? What if the tension between these ideas lies not in the evidence itself but in the context we use to interpret it?

To address this, it's important to clarify what I mean by "Creation" and "Creator." I do not, of course, refer here to the literalist interpretations often associated with young-Earth Creationism, which outright reject evolution and insist on a rigid, scriptural account of the natural world. Instead, my view of Creation encompasses the understanding of an intentional force—an underlying intelligence or purpose—that set the stage for the evolutionary processes we observe today. In this context, evolution and Creation are not adversaries but complementary parts of a greater whole.

This perspective came into sharp focus for me when I began to consider the remarkable complexity of DNA. Deoxyribonucleic acid, the molecule of life, is one of the most intricate and awe-inspiring systems we have ever encountered. DNA is the blueprint for all living organisms, containing the instructions for building and maintaining every cell, tissue, and organ in the body.

So, while all life evolved, it was DNA that heralded in that evolution, the evolution of all life. And where did the complexity, the knowledge base, the computer-like, the factory-like apparatus that is DNA come from? Dr. James Tour is the renowned American chemist and nanotechnologist known for his groundbreaking work in the fields of organic synthesis, molecular electronics, and nanotechnology.

Tour said, "The complexity and interdependence of cellular systems render it highly improbable that DNA, or any form of functional life, could emerge spontaneously through undirected chemical processes." He has spoken extensively about the challenges of explaining the naturalistic origins of DNA and other biomolecules essential to life. His main argument is that DNA is not just a sequence of nucleotides but a highly ordered structure requiring precise chemical conditions for its formation.

Laboratory efforts to replicate or synthesize molecules often rely on controlled conditions, skilled intervention, and purified chemicals. These conditions would not exist in a prebiotic Earth scenario. Dr. Tour critiques the field of prebiotic chemistry for failing to demonstrate plausible pathways to create complex molecules like DNA in a natural, unguided environment. He highlights the gap between producing simple molecules (e.g., amino acids or nucleotides) and assembling them into functional, information-carrying polymers like DNA.

DNA serves as an information storage system for life. How is it possible information, which is abstract and immaterial, could arise from purely material processes? The sequencing of nucleotides in DNA must be specific to encode functional proteins, but random processes are unlikely to generate meaningful sequences. DNA requires enzymes and proteins to replicate and function, yet these enzymes and proteins are themselves encoded by DNA, creating a dependency loop that challenges the idea of DNA arising independently.

Additionally, the interdependence of cellular components suggests that such systems would need to appear simultaneously, posing challenges to explanations based on naturalistic mechanisms. Tour also critiques scientific literature for making speculative claims without providing experimental validation. He has repeatedly said: "We have no idea how the molecules needed for life could have formed, let alone assembled into functional systems."

This underscores the mathematical improbability of assembling even a single functional DNA molecule through random processes. The number of possible sequences vastly outnumbers those that would be biologically viable. The sheer amount of information encoded within the DNA of even the simplest organism is staggering. For instance, the human genome consists of approximately 3.2 billion base pairs, an immense dataset that orchestrates the development and function of an organism with breathtaking precision.

What's more remarkable is the language-like structure of DNA. It operates with a four-letter code—adenine (A), thymine (T), guanine (G), and cytosine (C)—that functions much like a highly sophisticated programming language. This code directs the synthesis of proteins, the workhorses of cellular life, which carry out virtually every function necessary for survival.

The fidelity of DNA replication and its capacity to repair errors highlight an astonishing level of precision. Yet, DNA is also flexible, capable of mutations and recombination's that drive the diversity and adaptability central to evolution. As an atheist, looking at DNA, I marveled at the elegance of this system, viewing it as the ultimate testament to natural selection's power to shape complexity from simplicity. But some years ago,

I found myself asking the question: Did I view DNA and evolution this way because of the influence of my education and superficial science, which was again reinforced by atheist thinkers?

Now, based on the evidence, I see that DNA is far too complex to have evolved on its own. The intricate coding of DNA and the molecular machinery that translates and expresses it cannot simply be the result of undirected processes. This complexity itself is evidence of an intentional design—a kind of divine signature woven into the fabric of life.

To truly appreciate why the origin of DNA strains naturalism, we must look at the grand stage upon which it arrived. The biological code of life did not appear in a vacuum; it is the culmination of a cosmic narrative that began with the Big Bang. If we trace the lineage of a single carbon or phosphorus atom inside a DNA molecule, we find its origin story written in the primordial fires of the early universe and the intense gravitational furnaces of collapsing stars. The laws of physics beautifully explain how an exploding supernova scatters these raw elements across the dark expanse of space. Yet, those same physical laws are utterly silent when asked how those scattered atoms arranged themselves into a self-correcting, digital operating system. There is a profound, insurmountable gulf between forging raw matter and organizing that matter into a functional linguistic script.

What we must confront here is that the layout of DNA is not a problem of chemistry, but of information. Consider a computer program: the software can be saved onto a physical hard drive, but the metal and plastic of the drive did not create the code. The arrangement of data is an expression of mind, completely independent of the material hosting it. If you write a sentence in the sand on a beach, the message does not originate from the chemical properties of the silicon grains; it comes from an external intelligence. In DNA, the chemical bases—A, T, C, and G—are merely the physical ink and paper. The instructions they spell out are abstract, immaterial data. To claim that the laws of unguided chemistry generated the digital library of the genome is to mistake the physical properties of the ink for the mind of the author.

Based on the facts and complexity of DNA, and ironically, even based on many of those who proclaim it simply evolved naturally, they have to admit they simply cannot know how it evolved. They believe it evolved because they reject a Creation, and naturally, if they believe there was no Creation, then DNA would have had to evolve naturally.

The most telling validation of this mystery comes from the very secular scientists who would otherwise oppose my conclusions. When pushed past the glossy summaries found in popular textbooks, the world's leading evolutionary biologists and prebiotic chemists readily admit that they are utterly in the dark regarding the origin of the first self-replicating strands of DNA.

They will fiercely reject the idea of an intelligent Creator, yet in the same breath, they must concede that science currently possesses no viable, step-by-step naturalistic mechanism to explain how digital code emerged from a primordial soup. They operate on a foundational assumption that a material explanation *must* exist, but as a matter of raw empirical fact, they are forced to confess an absolute, collective ignorance as to how it actually happened.

To broaden this discussion, and to bring us back to my assertions about what the Big Bang offered, in the beginning, we must look beyond life and evolution to the origins of the universe itself. Evolution, after all, is not an isolated phenomenon. It exists within a larger framework, one that began with a moment of unparalleled significance: the Big Bang. This event, roughly 13.8 billion years ago, marks the beginning of time and space as we know them.

The singularity prior to the Big Bang is an enigmatic concept. It represents a point where all known laws of physics break down, a realm of infinite density and temperature. Within this singularity lay the potential for everything: the laws of physics, the fundamental forces, the particles that would form stars, galaxies, and eventually, life itself. It is, in many ways, the ultimate seed of Creation—a moment of extraordinary potential waiting to unfold.

As the universe expanded, the initial simplicity of the singularity gave way to increasing complexity. Subatomic particles coalesced into atoms; atoms gathered into stars; stars forged heavier elements through nuclear fusion. These elements were scattered across the cosmos by supernovae, enriching the interstellar medium with the raw materials for planets, oceans, and life.

Over billions of years, the universe transitioned from a chaotic, primordial state to one of astonishing order and diversity. This progression mirrors, on a cosmic scale, the process of evolution itself—a movement from simplicity to complexity, driven by inherent potential. Here, the question arises: what set this process in motion? Was the singularity's potential an accident, a fortuitous result of randomness? Or could it reflect the fingerprints of a Creator—a deliberate act that imbued the universe with the conditions necessary for life?

For me, the answer to this question deepens the wonder of existence. The fact that the universe operates according to precise laws, with constants finely tuned to allow for the emergence of stars, planets, and life, suggests not just randomness but intentionality.

Consider the relationship between the Big Bang and evolution. Evolution is often viewed as a purely biological process, but its roots extend to the very beginning of time. The elements that form DNA—the carbon, hydrogen, oxygen, nitrogen, and phosphorus—were forged in the hearts of stars billions of years before life emerged.

The processes that govern natural selection, adaptation, and biodiversity are built upon the physical laws established in the aftermath of the Big Bang. In this sense, evolution is not merely a biological phenomenon but a continuation of the universe's unfolding complexity—a manifestation of the singularity's potential. From this perspective, evolution becomes part of a grander narrative. It is not an isolated mechanism but a chapter in the story of a universe that began with a burst of energy and has since unfolded into galaxies, ecosystems, and sentient beings capable of pondering their origins.

The singularity's potential, realized through the Big Bang, set the stage for everything that followed, including the evolutionary processes that shaped life on Earth. Evolution remains a robust explanation for the diversity of life, and the Big Bang provides a well-supported model for the origins of the universe. Science explains much, but for all its explanatory power, does not answer the deeper "why" questions.

Why does the universe exist at all? Why does it operate according to intelligible laws? Why is it imbued with such profound potential? For me, the answer is a Creation and a Creator—a force or intelligence that set the universe in motion, embedding within it the potential for stars, planets, DNA, and consciousness. So, evolution itself should be seen as part of the Creator's design, the mechanism through which the universe's potential for developing life is realized over time.

In terms of the generalization of evolution, one does not have to consider if evolution is a fact, because based on the science, we and all living matter did evolve to be what we are today. Evolution, in almost every aspect, especially of life on Earth, is as clear as the summer sun, to use a line from Shakespeare's Henry the Fifth.

Also, the quote on top of the opening page of this chapter by Carl Sagan about us all being made in collapsing stars is accurate and is rooted in scientific fact. This idea captures the profound connection between human beings and the cosmos. It is a humbling reminder that every atom in our bodies has an ancient and awe-inspiring history tied to the life cycles of stars.

Let's dive deeper into how this connection is possible and what it reveals about our place in the universe. All the elements heavier than hydrogen and helium were created in the cores of stars through nuclear fusion. When massive stars reached the end of their lives, they exploded in supernovae, scattering these elements across the universe. Over billions of years, these elements became part of new stars, planets, and eventually, living beings like us.

For life, mutations are the source of genetic variation, the raw material for evolution. While most mutations are neutral or deleterious, some can confer benefits under specific conditions. Mutation rates vary between species and even among genes, adding another layer of complexity. Gene flow, or the exchange of genes between populations, introduces genetic diversity and can prevent speciation or lead to new evolutionary pathways. It highlights the interconnectedness of populations and the dynamic nature of gene pools.

Unlike natural selection, which emphasizes traits that enhance survival, sexual selection focuses on traits that improve mating success, such as elaborate plumage or behavioral displays. These traits may seem counterintuitive in terms of survival but are crucial to understanding evolutionary outcomes. While scientific principles explain how life evolved from the stars, this reductionist perspective focuses only on fundamental components, missing the broader picture: the deeper "how" and "why" behind evolution.

For example, if the Big Bang—initiating everything, including evolution—occurred naturally without any input from a Creator,

I would find it deeply surprising. As an atheist years ago, this idea seemed perfectly logical, but my deeper understanding of the universe's complexity now makes me question it. The central issue is the origin of DNA, the foundation of life. DNA introduces variation through random changes, allowing traits that improve survival and reproduction to become more common over generations.

But where did DNA itself come from? This is the core question. I contend that DNA could not have evolved naturally. It is more than just a chemical structure; it is a highly advanced code, complete with redundancy and error-correcting mechanisms—a "blueprint" for life that demands explanation beyond mere chance. The origin of this information presents a profound mystery. How did the first self-replicating molecules acquire the necessary information to kick-start life? The idea that such a system could arise from purely random chemical reactions strains credulity.

The philosopher Antony Flew, once a staunch atheist, cited the complexity and apparent design of DNA as a reason for his eventual belief in a Creation. Let's look deeper, DNA serves as the foundational blueprint for all living organisms, providing detailed instructions for constructing and maintaining life. Comparable to the schematic designs of a sophisticated factory,

DNA's genetic code directs the synthesis of proteins—the essential building blocks of life.

The sequences dictate growth, function, and development. Just as a factory's blueprints specify every step in production, DNA meticulously orchestrates the processes that sustain life. Despite decades of research, the origin of DNA's intricate information remains one of the most profound mysteries in science. The fact is, scientists are still uncertain about how life's foundational molecules arose or how the genetic instructions encoded within DNA came into existence.

This enigma challenges existing theories and underscores the need for innovative approaches to understanding the origins of life. While evolutionary theory explains many aspects of biological diversity, it struggles to address macroevolutionary events, such as the emergence of entirely new body plans or the origin of life itself. Critiques of current models highlight the limitations in explaining the mechanisms behind large-scale changes, suggesting these processes remain elusive and may extend beyond the scope of existing scientific paradigms.

Such observations prompt a reassessment of evolutionary frameworks and encourage openness to alternative perspectives. Also, a significant gap in evolutionary understanding lies in the functional differences between modern human brains and those of other hominids. Anatomical studies provide clues, but the biochemical and genetic factors behind advanced human cognition remain largely unexplained.

DNA's organization reflects an extraordinary level of complexity. Its hierarchical structure integrates genes, regulatory elements, and functional regions that coordinate cellular processes. Feedback loops allow DNA to regulate itself, ensuring stable yet adaptable cellular systems. This intricate interplay resembles computational networks, where inputs and outputs are precisely managed to maintain functionality.

Another one of DNA's most remarkable features is its ability to execute multiple processes simultaneously. This parallel processing enables cells to perform diverse functions—such as growth, repair, and adaptation—efficiently and in real time. This is not unlike advanced computing systems capable of running multiple complex algorithms concurrently, underscoring DNA's unparalleled efficiency.

The step-by-step mechanisms by which DNA is "read" and "executed" mirror the algorithmic functions of computer systems. Genetic instructions guide the synthesis of proteins, while built-in error correction mechanisms detect and repair replication errors. These self-correcting processes ensure the accuracy and stability of genetic information, akin to the error-checking protocols in computer systems that maintain data integrity.

Dr. Stephen Meyer is a prominent philosopher of science, author, and advocate for intelligent design (ID). He is known for his arguments that certain features of the natural world, particularly the complexity of biological information, provide evidence for an intelligent cause behind life and the universe. Meyer argues that the complexity and specified information in biological systems, particularly DNA, are best explained by the action of an intelligent cause rather than undirected natural processes like Darwinian evolution.

Dr. Meyer further argues that DNA's processes resemble the execution of computer programs. The genetic code, through transcription and translation, converts nucleotide sequences into functional proteins with algorithmic precision. This analogy underscores DNA's role as a sophisticated information-processing system, raising profound questions about its origins and the potential for intelligent design.

DNA's mechanisms for ensuring genetic fidelity are akin to the quality control systems in factories. During replication, proofreading enzymes correct errors, while repair systems address damage caused by environmental factors or cellular stress. These maintenance processes uphold genome integrity, much like factories rely on inspections and repairs to sustain production standards. Just as factories use management systems to oversee production, DNA employs regulatory elements such as promoters and silencers to control gene expression.

These elements enable cells to respond to environmental cues and developmental needs, ensuring resources are used efficiently. This adaptability mirrors the flexibility of modern factories, which adjust production to meet changing demands. A factory's sustainability depends on effective waste management and timely maintenance. Similarly, DNA ensures long-term stability through robust repair mechanisms and processes like apoptosis, which eliminates damaged cells. These systems prevent errors from accumulating, maintaining the organism's health and functionality over its lifetime.

The intricate design and functionality of DNA challenge conventional explanations of life's origins. Its hierarchical organization, error-correcting algorithms, and dynamic adaptability reveal a system of unmatched sophistication. For Drs. Tour and Meyer, the study of DNA goes beyond biology, inviting philosophical and scientific exploration of the profound intelligence that may underlie life's design.

The "appearance of design" is so compelling that even staunchly atheistic scientists, such as Richard Dawkins, acknowledge it in their work. Dawkins describes biology as "the study of complicated things that give the appearance of having been designed for a purpose." The theory of evolution is remarkably accurate even as we discover more, and provides a robust framework for understanding the development of life, it does not address the ultimate questions of existence: Why is there something rather than nothing? What is the source of the laws of nature? Why does the universe permit life?

It is essential to emphasize that the idea of outside influence in evolution does not necessarily conflict with scientific inquiry. Many scientists, including prominent figures like Francis Collins, the director of the Human Genome Project, see no contradiction between evolution and belief in a Creation. Collins describes the process of evolution as a tool used by the Creator to bring about the diversity of life.

The "origin of life" is best described as the chemical and physical processes that brought into existence the first self-replicating molecule. It differs from the "evolution of life" because Darwinian evolution employs mutation and natural selection to change organisms, which requires reproduction. Since there was no reproduction before the first life, no "mutation-selection" mechanism was operating to build complexity. Hence, origin theories cannot rely upon natural selection to increase complexity and must create the first life using only the laws of chemistry and physics.

There are so many problems with purely natural explanations for the chemical origin of life on earth that many scientists have already abandoned all hopes that life had a natural origin on earth. Even the staunch atheist Richard Dawkins, in his book

The Blind Watchmaker (1986), makes a striking observation: "Physics books may be complicated, but the objects and phenomena that a physics book describes are simpler than a single cell in the body of its author. The author, in turn, consists of trillions of cells, many of them different from each other, organized with intricate architecture and precision engineering into a working machine capable of writing a book. ... Each nucleus ... contains a digitally coded database larger, in information content, than all 30 volumes of the Encyclopaedia Britannica put together."

In this passage, Dawkins discusses the astounding complexity of biological systems and contrasts them with the comparatively simpler laws of physics. He underscores the remarkable intricacy of life at the cellular level, particularly highlighting the vast information content encoded in DNA. DNA's double-helix structure, discovered by James Watson and Francis Crick in 1953, is an elegant and efficient design. DNA is extraordinarily compact. A single gram of DNA can theoretically store 215 petabytes of data—far beyond the capacity of any current artificial storage device. This level of efficiency suggests an advanced system of information management.

Certain biological systems, such as those controlled by DNA, exhibit irreducible complexity, meaning that they cannot function if any part is removed or altered. For example, the processes of transcription (DNA to RNA) and translation (RNA to protein) rely on a vast network of interdependent molecules. The simultaneous existence of these components is essential, complicating piecemeal evolutionary explanations.

Even Bill Gates famously compared DNA to a computer program, stating "DNA is like a computer program, but far, far more advanced than any software ever created." If advanced human systems require intelligent design, does DNA's superiority not imply the same? DNA relies on proteins for its replication and function, but proteins are synthesized based on instructions encoded in DNA.

This interdependence creates a chicken-and-egg problem: which came first? The simultaneous emergence of DNA and proteins seems improbable without guidance or an overarching plan. In information theory, all codes and languages originate from an intelligent mind. DNA, the most advanced information system known, exemplifies this principle.

Random processes have never been observed to generate coherent, functional codes, highlighting the evidence for design. The universality of the genetic code implies a common Designer who created all life forms with a shared blueprint, reflecting a purposeful and harmonious Creation. From its precise structure to its information-rich coding system, DNA's intricate design and unparalleled efficiency challenge naturalistic explanations, pointing instead to the involvement of an intelligent Creator.

My deeper understanding of evolution, particularly in relation to the intricate complexities of DNA, has led me to conclude that both design and Creation are integral to the origins of life. The staggering precision and interdependence of biological systems, especially at the molecular level, suggest a purposefulness that goes beyond mere chance.

Reflecting on my formative years as an atheist, I see now how my earlier, more superficial interpretations of science led me to a simplistic and, ultimately, incomplete conclusion: that the fact of evolution necessarily precluded the possibility of Creation. This notion, which once seemed self-evident, I now recognize as deeply flawed. Evolution, as a process, does not negate Creation but may instead point to an underlying design—a harmony and intent woven into the very fabric of life itself.

"The fine-tuning of the universe for intelligent life is so improbable that it cries out for an explanation."

William Lane Craig

"The laws of physics seem peculiarly suited to the existence of life. If any one of about 20 physical constants were 'off' by even a small amount, the universe could not sustain life as we know it."

Leonard Susskind

CHAPTER 5

Fine Tuning of the Universe

The existence of the universe and its intricacies is a profound mystery that continues to captivate humanity. One approach to exploring this mystery is through the principle of causality, the logical idea that every effect has a cause, and that nothing comes from nothing. However, this principle presents a paradox: if the universe had a cause, then what caused that cause? While some posit an infinite regress of causes, such explanations fail to account for the ultimate origin of the universe we now see. Importantly, for the purposes of this chapter, and indeed this book, I put aside speculation about what preceded the singularity or the Big Bang because it changes nothing in terms of the Creation event at the Big Bang which is my focus. Recent scientific advancements suggest the possibility of other Big Bangs previous to the one we now see, not just the one that birthed our universe. While these ideas hold significant implications, they do not change the focus of this book. Whether our universe was part of a broader multiverse, other previous Big Bangs or the result of a single unique event, the Big Bang remains a definitive beginning point for our observable universe.

At the singularity, where time and space as we understand them cease to exist, all known physical laws break down. The universe transitions from a state of infinite density and no dimensions to one of expansive time and space. This moment, often referred to as the "new beginning," transcends human comprehension and resonates with the reflections of some of the greatest minds in science. Stephen Hawking, who extensively explored singularities, observed, "The universe would not be arbitrary but would be determined by a set of laws. These laws may have originally been decreed by God, but it seems that He has since left the universe to evolve according to them." The singularity, whether viewed as the result of one Big Bang or part of a larger sequence, underscores the profound complexity and mystery of the universe's origins. This leads naturally to the concept of fine-tuning—the precise calibration of physical constants and laws that govern the universe.

Fine-tuning refers to the extraordinary specificity of these constants, which, if altered even slightly, would render the universe incapable of supporting life. This apparent precision has not gone unnoticed by leading scientists. Physicist and cosmologist Paul Davies remarked, "There is for me powerful evidence that there is something going on behind it all... The impression of design is overwhelming." Similarly, Stephen Hawking also observed, "The laws of science, as we know them at present, contain many fundamental numbers... The remarkable fact is that the values of these numbers seem to have been finely adjusted to make possible the development of life." Astrophysicist Fred Hoyle famously stated, "A common-sense interpretation of the facts suggests that a super intellect has monkeyed with physics, as well as with chemistry and biology, and that there are no blind forces worth speaking about in nature." These reflections align with the idea that the universe's fine-tuning is not a product of random chance but a phenomenon that suggests purposeful design. Even scientists who refrain from drawing Creation event conclusions recognize the problem of fine-tuning.

Theoretical physicist Alan Guth once remarked, "The apparent fine-tuning of the constants of nature to support life is a problem that is recognized by all physicists." This statement captures a profound and widely acknowledged mystery at the heart of modern physics and cosmology. What Guth refers to is the observation that the fundamental constants of nature—the values that govern the behavior of forces, particles, and the structure of the universe—appear to be calibrated with astonishing precision to allow for the emergence of life. If any of these constants were even slightly different—by just a fraction of a percent—the universe as we know it would be inhospitable to life. Stars might not form, chemistry might not function, and complex structures like galaxies, planets, or living organisms could never emerge. This extraordinary balance has prompted many physicists and philosophers to ask: why do the constants have the values they do? Is it mere coincidence, a necessity of deeper laws we don't yet understand, or evidence of design?

Guth's comment reflects the fact that, regardless of one's interpretation—naturalistic, multiverse, or theistic—the fine-tuning problem is not speculative or theological in nature; it is empirical. The constants are what they are, and the probability of their life-permitting range appearing by chance, in a single-universe scenario, is vanishingly small. Once again, we must reflect on the Singularity and the Big Bang, which initially offered potential. Yet, it is crucial to recognize that this was not a "passive" potential.

The true beginning offered only potential, but that potential, in essence, caused our reality. It was a loaded potential, containing within it the specific mathematical instructions and physical blueprints required to bring forth a structured cosmos. This potential did not simply remain abstract; it acted as the invisible architect that unfolded into a universe remarkably fine-tuned, where the precise conditions necessary for life and complexity were realized against seemingly insurmountable odds. In this sense, the potential was the seed, and the reality we inhabit is the inevitable bloom of a perfectly designed core.

The fine-tuning of the universe manifests in numerous ways, from the fundamental constants of nature to the specific conditions that make life possible. The constants of the universe are like the fixed rules or settings that govern how everything works, from the smallest particles to the largest galaxies. These values don't change and are the same everywhere in the universe. They make sure that the universe runs smoothly and predictably. Without these constants, the universe wouldn't look anything like it does now—stars, planets, and even life might not exist. The cosmological constant controls how fast the universe is expanding—too fast, and galaxies wouldn't form; too slow, and everything might collapse. Even at the tiniest level, constants like the charge of an electron and the mass ratio between protons and electrons allow atoms to stick together and form the molecules that make up life.

It's as if the universe is perfectly "fine-tuned" to support life. Scientists and philosophers often wonder why these values are exactly what they are. Some suggest it's just a lucky coincidence; others think it's due to the laws of nature, however if looked at objectively, it points to the work of a Creator who designed the universe with care. Consider the following: The cosmological constant is a term in Einstein's equations of general relativity that represents a kind of energy that fills space itself. In simple terms, it's like a "push" that makes the universe expand faster over time, counteracting the pull of gravity. To visualize it: imagine you're blowing up a balloon. Normally, you'd expect the balloon to expand slower as it gets bigger because the rubber is stretched tight. But if there were an invisible force constantly inflating the balloon faster and faster, that's kind of what the cosmological constant does to the universe.

The cosmological constant is a number that represents the energy density of empty space, or "vacuum energy." This value is associated with the cosmological constant, introduced by Albert Einstein in his equations of General Relativity. This cosmological constant is a very, very tiny number that shows how much energy in in empty space. In simple terms, it's like a built-in property of space itself that makes space expand. Even in places where there's no matter or light, this energy is still there, pushing the universe to grow larger. Scientists associate this with "dark energy," a mysterious force causing the expansion of the universe to speed up rather than slow down. Without this, the universe's behavior and growth wouldn't match what we observe today.

In terms of fine tuning, if the expansion constant were even slightly different, the universe would either collapse in on itself or expand too rapidly for galaxies and stars to form. The observed value of the cosmological constant, which governs the energy density of empty space and the expansion rate of the universe, is one of the most finely tuned quantities in physics. Scientists estimate its precision to be so extreme that it has been described as accurate to "one part in a trillion trillion trillion trillion trillion trillion trillion trillion trillion trillion." To put this staggering level of precision into perspective, imagine firing an arrow at a target located a trillion light-years away—an unfathomable distance—and somehow hitting the exact bullseye. This analogy illustrates just how delicate the balance of this value must be for the universe to exist as it does.

If the cosmological constant were even slightly larger, the universe would have expanded so rapidly after the Big Bang that galaxies, stars, and planets would never have formed. On the other hand, if it were slightly smaller, the universe could have collapsed back in on itself, preventing the development of complex structures and life. The precision required for this constant to fall within the narrow range that allows for a stable and life-supporting universe is so extraordinary that it challenges our intuition about randomness or chance. This remarkable fine-tuning raises profound questions about the nature of the universe. Is this level of precision a mere coincidence, the result of some unknown physical necessity, or evidence of intentional design? The extraordinary fine-tuning of the cosmological constant continues to be one of the greatest mysteries in modern science, inspiring debates that stretch across physics and philosophy.

The gravitational constant is a fundamental number in physics that determines the strength of gravity, the force that attracts objects with mass toward one another. While its precise value is extremely small, its impact on the universe is profound.

If gravity were even slightly stronger, gravity would be more intense, causing stars to burn through their fuel much faster. Stars rely on a delicate balance between gravity pulling inward and the pressure from nuclear fusion pushing outward. A stronger gravitational constant would compress stars more tightly, increasing the rate of fusion. This would shorten the lifespan of stars, leaving insufficient time for planets to form and for life to evolve.

Conversely, if the strength of gravity were weaker, gravity would not be strong enough to trigger nuclear fusion in stars. Fusion is the process that powers stars and produces the elements necessary for life, like carbon and oxygen. Without it, stars would fail to ignite, and the universe would remain cold and dark, with no light or heat to sustain life. This delicate balance in the value of the gravitational force also affects the formation of galaxies, stars, and planets. If gravity were slightly stronger or weaker, galaxies might collapse too quickly or fail to form altogether. The formation of planetary systems, like our solar system, depends on just the right amount of gravitational pull to gather dust and gas into planets while allowing them to orbit stars stably. In essence, the precise value of gravity is one of the "fine-tuned" constants in the universe, enabling the cosmic environment necessary for life to exist. This remarkable balance has led many scientists to ponder whether the universe's physical laws are uniquely suited to support life, reinforcing the fine-tuning argument.

The electromagnetic force governs the interactions between charged particles. Its strength determines the stability of atoms and the ability of molecules to form. If this force were slightly stronger, chemical bonds would not form correctly. If it were weaker, atoms would not hold together. The proton-to-electron mass ratio is approximately eighteen hundred to one. This specific ratio is critical for the stability of atoms. Even a slight variation would destabilize matter and prevent the formation of the complex molecules necessary for life. Earth itself resides in the "Goldilocks zone" of our solar system, where conditions are "just right" for life. If Earth were slightly closer to or farther from the Sun, its surface temperature would either be too hot for liquid water or too cold to sustain it. The presence of a stable star, a magnetic field, and an atmosphere further underscores the precise conditions that make life possible.

Even water, one of the most basic necessities for life, is astounding and in terms of fine tuning it is awe-inspiring. Water is one of the most remarkable substances in the universe, and its unique properties make it essential for life as we know it. Despite its simplicity—just two hydrogen atoms bonded to one oxygen atom—water exhibits a range of characteristics that are unmatched by almost any other compound. These properties, rooted in its molecular structure, create the conditions necessary for life to exist and thrive. Water's ability to dissolve a wide variety of substances is one of its most important features. This is due to its polar molecular structure, with a slightly positive charge on the hydrogen atoms and a slightly negative charge on the oxygen atom. This polarity allows water molecules to interact with and break apart other molecules, effectively dissolving them. In biological systems, water acts as the medium in which nutrients, minerals, and gases like oxygen are transported. For example, blood is largely water and carries essential substances to cells while removing waste products.

This property also facilitates chemical reactions. In living organisms, water dissolves reactants and provides the environment where complex biochemical processes occur, such as the breakdown of food during digestion and the replication of DNA. Water has a high specific heat capacity, meaning it can absorb or release a significant amount of heat with only a small change in temperature. This property is crucial for maintaining stable environments. On a global scale, water bodies like oceans and lakes moderate Earth's climate by absorbing heat during the day and releasing it slowly at night. In living organisms, water plays a vital role in regulating body temperature. For example, sweat, which is primarily water, evaporates from the skin, carrying away heat and cooling the body. Water is not just a passive medium for reactions—it actively participates in many of them. For example: Hydrolysis reactions, where water breaks down complex molecules like proteins and carbohydrates into simpler units that cells can use. In photosynthesis, water molecules are split into oxygen, protons, and electrons, providing the energy that fuels nearly all life on Earth.

Water remains liquid across a broad temperature range (0 Celsius to 100 Celsius at standard pressure), which is critical for life. This range is wide compared to most other substances and allows water to exist in liquid form in many environments on Earth. Liquid water is necessary for life because it acts as a solvent and medium for biochemical processes. Without it, cells—tiny factories of life—would not function. The solid (ice) and gaseous (steam) forms of water also play roles in the planet's climate and ecosystems, but liquid water is the foundation of life. The Density Anomaly unlike most substances, water expands when it freezes. This makes ice less dense than liquid water, allowing it to float. This property is vital for aquatic ecosystems. When ice forms on the surface of lakes or oceans, it insulates the water below, preventing it from freezing solid and allowing life to survive during harsh winters. If ice sank, water bodies would freeze from the bottom up, making it nearly impossible for aquatic life to persist.

Water has a high surface tension, meaning its molecules stick together strongly due to hydrogen bonding. This cohesion enables water to form droplets and allows it to move through narrow spaces, such as the tiny capillaries in plants and animals. In plants, water's cohesive and adhesive properties allow it to be drawn up from the roots to the leaves against gravity—a process known as capillary action, which is crucial for transporting nutrients. Water is transparent to visible light, allowing sunlight to penetrate oceans, lakes, and rivers. This property is essential for photosynthesis in aquatic plants and algae, which forms the base of nearly all food chains. Transparency also allows marine life to navigate and thrive in underwater environments.

Water's role in supporting life extends to its interaction with biological molecules. It stabilizes proteins, DNA, and other complex molecules, helping them maintain the shapes necessary for their functions. Water also enables the formation of cell membranes by interacting with lipid molecules, creating the boundary that separates the interior of a cell from its environment. So even water is far more than just a common liquid—it is the lifeblood of our planet and an essential ingredient for life. Its ability to dissolve substances, regulate temperature, facilitate chemical reactions, and remain liquid over a wide range of conditions makes it uniquely suited to support life. The incredible properties of water are so finely tuned that they have led many scientists to marvel at the role it plays in sustaining life on Earth. Without water, life as we know it would be impossible, highlighting its fundamental importance in the story of life's existence and evolution.

Dark matter and dark energy, though not directly observable, play crucial roles in the universe's structure and expansion. Dark matter provides the gravitational scaffolding for galaxies, while dark energy drives the accelerated expansion of the universe. If their proportions were even slightly different, the cosmos would be inhospitable to life. The sheer improbability of these precise conditions occurring by chance has led many to conclude that the universe is the result of intentional design. William Lane Craig, a philosopher and theologian, argues that the fine-tuning of the universe points to an intelligent designer. He explains, "The odds of a life-permitting universe occurring by chance are so astronomically low that it is more plausible to believe in a purposeful designer." The concept of a designer also aligns with the anthropic principle, which states that the universe's laws and constants appear fine-tuned to allow the existence of observers. Hugh Ross, an astrophysicist, writes, "The fine-tuning of the universe is so exact that it's as if the universe knew we were coming."

The evidence for fine-tuning extends beyond physics into the realms of chemistry and biology. The properties of carbon, for instance, make it uniquely suited to form the complex molecules required for life. Carbon's ability to form long chains and rings underpins the chemistry of life, from DNA to proteins. Furthermore, the conditions on Earth that allow for the emergence and sustenance of life are remarkable. The planet's distance from the Sun, its axial tilt, and the presence of a large moon to stabilize its rotation are just a few examples of the factors that make Earth uniquely habitable. The fine-tuning of the universe compels us to confront profound questions about its origins. While some argue that fine-tuning can be explained through naturalistic mechanisms or multiverse theories, such explanations often rely on speculative assumptions. The idea of a designer, by contrast, provides a coherent and intuitive explanation for the universe's extraordinary precision.

As we continue to uncover the intricacies of the cosmos, the evidence for fine-tuning grows ever more compelling. Whether viewed through the lens of physics, chemistry, or biology, the universe's remarkable suitability for life points to a reality that transcends mere chance. It invites us to consider the possibility of a designer who orchestrated the cosmos with intention and purpose. The question that arises from these observations is: how did the universe come to be so finely tuned? Was it simply a coincidence, or is there a more profound explanation? Even if one doesn't believe in a Creation, the question remains: how did everything come into existence from nothing? If we assume that the universe indeed came from nothing, we are faced with a multitude of unanswered questions. What exactly is "nothing"?

Is it a state of complete absence of matter, energy, and space? Or is it simply a label we use to describe something that is beyond our comprehension?

Furthermore, if the universe came from nothing, then how do we explain the laws of physics that govern its behavior? Where did these laws come from? Did they emerge spontaneously, along with the universe, or were they somehow inherent in the fabric of nothingness? The idea that the universe came from nothing is often seen as a more "scientific" explanation than the idea of a Creation. However, this perspective assumes that the concept of nothing is somehow more rational or easier to understand than the concept of a Creation. In reality, the idea that the universe came from nothing is just as mind-boggling as the idea of a Creation. In fact, I would suggest that the concept of nothing is even more incredible than the concept of a Creation. At least with the concept of a Creation, we have some framework for understanding how the universe might have come into existence.

I pointed out in my book's introduction, the evidence I will present leads to a compelling conclusion: a Creator or Creator's designed our universe, and the Big Bang marks a definitive Creation event including the inherent potential for the fine tuning. The vastness of the cosmos, the precision of its laws, and the intricate tapestry of life all point toward intentionality rather than chance. However, while the evidence suggests the "what" of Creation—a universe brought into being with remarkable order and purpose—the "how" and "why" of this Creation remain, at least for now, beyond our reach. Perhaps this unknowability is by design, an invitation to embrace both curiosity and humility in equal measure. It is not a weakness to admit that the purpose of Creation lies outside our current understanding; rather, it is an acknowledgment of the profound and infinite nature of the question itself. The journey to understand Creation is as much about the search as it is about the answers.

"If quantum mechanics hasn't profoundly shocked you, you haven't understood it yet."

Niels Bohr

"The atoms or elementary particles themselves are not real; they form a world of potentialities or possibilities rather than one of things or facts."

Werner Heisenberg

CHAPTER 6

Quantum Mechanics and a Creation

The quotes above capture the essence of quantum mechanics as a field that challenges our understanding of reality, inspires awe, and provokes deep reflection. My journey into the fascinating world of Quantum Mechanics began after watching a simple yet transformative video illustration of the Double Slit Experiment. That video was more than just an explanation; it was a spark that ignited a journey of discovery, profoundly shaping my view. The realization that, in the quantum realm, reality defies our everyday expectations was both enlightening and deeply inspiring. It opened the door to exploring questions about existence, reality, and even Creation. Quantum mechanics doesn't just challenge our understanding—it redefines the very foundation of what we believe about reality. It introduces phenomena that seem paradoxical—particles existing in multiple states simultaneously, nonlocality where particles influence each other instantaneously regardless of distance, and the strange interplay between observation and reality. These phenomena are not mere curiosities; they strike at the heart of what we mean by "existence" itself.

Among the many astonishing concepts I encountered, three stood out as particularly revolutionary in shaping my considerations about the nature of existence and the evidence of a Creation event: the Double Slit Experiment, Time Dilation, and Quantum Entanglement. Each provides unique insight into the universe's inner workings, and collectively, they suggest a cosmos of precision, interconnectedness, and intentional design. Quantum mechanics is an immense and intricate field, brimming with concepts that challenge intuition. Beyond these three, numerous other ideas expand our understanding of reality. Quantum Superposition shows that a particle can exist in multiple states or locations simultaneously until it is observed or measured. Superposition exemplifies the universe's fundamental indeterminacy at the microscopic scale. Quantum Tunneling describes how particles can pass through energy barriers they classically shouldn't surmount, such as electrons "tunneling" through potential walls. This phenomenon underpins technologies from semiconductors to nuclear fusion in stars, showing that the quantum world operates beyond classical logic.

The Uncertainty Principle, formulated by Werner Heisenberg, asserts that certain pairs of properties—like position and momentum—cannot be simultaneously measured with absolute precision. This isn't a limitation of our tools but a fundamental property of nature. Quantum Decoherence explains why quantum systems, when interacting with their environment, appear to transition into classical behavior, providing insight into how the macroscopic world emerges from quantum laws. The Measurement Problem arises because observing a quantum system collapses it from multiple possibilities into a single outcome. Yet why observation triggers collapse remains unresolved—a profound mystery of science. Nonlocality demonstrates that quantum correlations appear instantly across vast distances, defying classical constraints of space and causality. Quantum Field Theory, the most advanced quantum framework, combines quantum mechanics and special relativity, describing particles as excitations in underlying fields and offering a unified perspective of the microscopic world. Physicists also aim to reconcile quantum mechanics with general relativity, seeking a "theory of everything" through approaches such as string theory or loop quantum gravity, striving toward the unification of physics.

With this context, we can now explore the three phenomena most relevant to understanding the universe's deep structure, beginning with the Double Slit Experiment. This is one of the most famous experiments in physics because it reveals how strange the microscopic world can be. It shows that tiny particles, such as electrons and photons, do not behave like ordinary objects. Instead, they can act like waves or like particles, depending on how the experiment is set up. Imagine a wall with two narrow slits and a screen behind it to record where particles land. If particles behaved like small solid objects, we would expect them to pass through one slit or the other and form two simple bands on the screen—similar to throwing marbles through two openings. What actually happens is very different. When nothing is placed at the slits to track the particles' paths, an interference pattern appears on the screen. This pattern looks like overlapping ripples on water and shows that each particle spreads out like a wave and passes through both slits at once, interfering with itself.

The pattern changes when a device is added to detect which slit each particle goes through. As soon as this path information is physically recorded, the interference pattern disappears. The particles now behave like individual objects, forming two broader bands on the screen. This change happens even if no person ever looks at the data. It is caused by the physical interaction between the particle and the measuring device, not by human awareness. Many popular explanations suggest that a human observer "collapses the wave" simply by watching the experiment. This is misleading. In physics, an observer does not mean a conscious mind—it means a measuring interaction. A camera, detector, or even the surrounding environment can destroy the interference pattern. The result is the same whether or not anyone is present to observe it. The experiment was first performed in 1801 by Thomas Young, who used light to demonstrate its wave nature. In the 20th century, scientists discovered that electrons also produce the same interference pattern, proving that wave-like behavior is a general feature of the quantum world. The experiment has since been repeated with many kinds of particles, always with the same result. The importance of the Double Slit Experiment lies in what it tells us about nature. At the quantum level, outcomes depend on physical interactions, not on human perception. While physicists continue to debate how measurement should be interpreted, it reveals a universe governed by precise but unfamiliar rules, where measurement is a physical process that shapes what can be observed.

Equally transformative is Time Dilation, a phenomenon from Einstein's theory of relativity that intersects intriguingly with quantum physics. Time dilation refers to the fact that time moves differently for observers in relative motion or within varying gravitational fields. There are two types of time dilation. Relative Velocity Time Dilation occurs as an object approaches the speed of light, causing time to slow relative to a stationary observer. Gravitational Time Dilation occurs when time slows in stronger gravitational fields compared to weaker ones. Evidence of time dilation is abundant. The Hafele-Keating Experiment of 1971 demonstrated that atomic clocks flown around Earth on commercial jets experienced time differently than stationary clocks, exactly as predicted. GPS technology also relies on corrections for both forms of time dilation to maintain accuracy; without them, locations would drift by kilometers each day. High-speed particles in accelerators live longer than they would at rest, confirming relativistic predictions.

Astronomical observations also show that light from stars near massive objects is redshifted due to slower time in strong gravitational fields, matching theoretical predictions. Time dilation reveals that time is not absolute but a relative dimension intertwined with space to form the spacetime continuum. It challenges our everyday sense of time as linear and universal, inviting us to rethink beginnings, endings, and the flow of existence. The Block Universe theory extends this idea: all moments—past, present, and future—exist simultaneously, and our perception of sequential time is an illusion. Einstein himself suggested that the distinction between past, present, and future is only a "stubbornly persistent illusion." Some physicists even speculate that time might emerge from particle interactions rather than existing independently, suggesting a deeper layer of reality. Time dilation and relativity reveal a universe structured with immense precision and complexity, raising metaphysical questions about who or what designed this temporal framework and why. Could this be another hint toward a purposeful

Creation, providing a stage for conscious beings to experience, learn, and grow?

This lead us to Quantum Entanglement, a phenomenon in which particles become inseparably linked so that the state of one instantly determines the state of the other, no matter the distance between them. Imagine two entangled electrons sent to opposite ends of the galaxy. Measuring the spin of one immediately reveals the spin of the other. Einstein famously called this "spooky action at a distance." Entanglement violates classical notions of locality and causality, suggesting an underlying unity to the universe. It has been repeatedly confirmed through experiments testing Bell's Theorem, demonstrating that no classical mechanism can explain the observed correlations. The implications of entanglement are profound. Reality is fundamentally nonlocal, and connections can exist instantaneously. Precision in quantum laws is necessary for entanglement to occur, suggesting fine-tuning. The interconnectedness of entangled particles parallels ideas of a purposeful Creation. Entanglement may metaphorically reflect a Creator's connection to all aspects of Creation, beyond the constraints of space and time.

Together with the Double Slit Experiment and Time Dilation, entanglement emphasizes the precision, interconnectedness, and depth of the universe—revealing a cosmos that is structured, responsive, and possibly intentional. Observation, time, and connection emerge not as arbitrary rules but as fundamental aspects of a coherent, intricately structured system. Quantum mechanics compels us to reconsider what it truly means for something to exist, how time flows, and how the universe's elements are entwined in a vast, interdependent web. It invites us to step beyond the familiar, material view of reality and recognize that the cosmos is more than a backdrop for random events—it may be an exquisitely crafted stage upon which existence unfolds with deliberate intent. The Double Slit Experiment, Time Dilation, and Quantum Entanglement reveal a universe of staggering sophistication, where patterns emerge that are simultaneously precise and mysterious. Particles that behave as waves until observed, time that stretches and contracts according to motion and gravity, and entangled particles whose states are linked across immense distances—all these phenomena point toward an underlying order that is far from accidental.

They suggest a universe that is deeply relational, where the observer and the observed are inseparable participants in the unfolding reality. The patterns of quantum behavior, the malleable rhythm of time, and the invisible threads of entanglement hint at a cosmos whose architecture is both intelligible and awe-inspiringly beautiful. These principles also illuminate the human experience in profound ways. We are not passive spectators in a disconnected universe; rather, our consciousness, curiosity, and perception are woven into the very fabric of reality. The act of observing, questioning, and seeking knowledge is itself a form of participation, a reflection of the universe's relational nature. In recognizing this, we glimpse a profound resonance between human awareness and cosmic structure, where the act of discovery mirrors the universe's own elegance and intentionality.

In witnessing the subtleties of quantum phenomena, the elasticity of time, and the mysterious bonds of entanglement, we encounter not only scientific truths but profound philosophical implications. These discoveries call us to humility, reminding us that even as we probe the smallest particles and the vastest stretches of spacetime, our understanding is always partial, our perspective finite. Yet within this limitation lies profound empowerment: the knowledge that we are capable of perceiving threads of order, glimpsing patterns, and intuiting the astonishing interconnectedness of all things. The universe, in this sense, becomes a mirror reflecting our own capacity to wonder, to comprehend, and to participate in its unfolding narrative. Contemplating the Double Slit Experiment, Time Dilation, and Quantum Entanglement reveals a reality that is simultaneously fragile and resilient, chaotic and ordered, fleeting and eternal. These phenomena provide a window into the architecture of existence itself, showing a tapestry in which every particle, moment, and event appears interwoven with subtle precision.

Observation, time, and entanglement are not isolated curiosities—they are signatures of a universe in which every element is meaningful and responsive. They hint at a cosmos where intelligence, order, and purpose may be inherent, and where the emergence of consciousness is woven into the very fabric of reality. The universe invites us to engage with it deeply, to marvel at its precision, to reflect on its beauty, and to recognize our place within it. In the subtle interplay of particles, the stretching and folding of time, and the invisible threads that link all matter, we discern a universe so remarkably orchestrated that it speaks to a mind behind its design—a Creator whose presence is suggested not only through the laws that govern reality but through the very act of our noticing, questioning, and awe. The cosmos, in all its complexity, interconnectedness, and elegance, becomes a living testament to possibility, meaning, and wonder—a silent, sublime invitation to explore, contemplate, and revere the astonishing unity of existence. In the elegance of quantum mechanics lies a profound question: could this be mere accident? The universe's subtle symmetries, its breathtaking complexity, and the strange, counterintuitive rules that govern reality certainly hint at something deeper. Something beyond randomness. This is utterly profound.

"What I see in Nature is a magnificent structure that we can comprehend only very imperfectly, and that must fill a thinking person with a feeling of humility."

Albert Einstein

"If we are in a simulation, then what we perceive as the laws of nature are really just the rules of the program."

Rizwan Virk

Chapter 7

Reflections on a Designed Universe

Upon review of the evidence presented in this book, it becomes clear that our universe had a deliberate, purposeful beginning. The Big Bang was a finely tuned occurrence governed by precise mathematical laws. The balance of physical constants, the complexity of DNA, and the structured nature of quantum mechanics all point to creation. While we may not fully comprehend the reason or mechanisms behind it, the evidence compels us to re-evaluate our understanding of existence. As we explore the mysteries of creation, the boundaries between science and philosophy blur. This intricate tapestry underscores the profound beauty and complexity of our universe. The fine-tuning of reality suggests not chaos, but order—so precise that even the slightest change would render life impossible. Such precision invites us to not only analyze but also appreciate the wonder of existence.

Looking back, my journey from atheism to Scientific Creationism reveals how a sincere, evidence-based pursuit of the facts can radically reshape what we believe about our origins.

The deeper I explored, the more I realized that randomness alone could not explain the extraordinary precision of our universe. Understanding creation is not just an intellectual pursuit but a shift in perspective, reshaping how we see the world and our place within it. Even as an atheist, I was inherently a positive thinker. Why wouldn't I be? Life itself is a wonder, and the world is truly beautiful. I marveled at nature's intricacies. My optimism was already a quiet acknowledgment of the extraordinary gift of existence. I share this now to emphasize that my change in belief about Creation was not born out of necessity—like someone searching for meaning—nor was it the result of a spiritual awakening.

Truth is, I've been fortunate to lead a deeply fulfilling life, rich with meaning and joy. I was blessed with a wonderful family and cherished grandchildren who filled my days with light and laughter, making every moment feel like a gift. Have I faced great challenges? Absolutely. Have I endured moments of despair or experienced failure? Of course! Yet, these experiences have been integral to my life, shaping who I am and deepening my appreciation for life's beauty and resilience. Professionally, I thrived by viewing challenges as opportunities for growth and staying actively engaged with the people around me. Whether navigating difficulties or celebrating successes, I found purpose in building connections, fostering collaboration, and maintaining a resilient spirit. These moments of triumph and trial have shaped me, allowing me to embrace life's imperfections with gratitude and grace.

Human connection has always been a cornerstone of my life. In every interaction—whether personal or professional—I made it a point to recognize and appreciate the good in others. This ability to see and celebrate admirable traits in those I encountered is something I've always taken pride in and which certainly helped my professional career. The truth is, during my years as an atheist, I experienced a profound sense of contentment and fulfillment within my worldview. There was no spiritual void or longing that needed to be addressed; instead, I found joy and purpose in the tangible and the present. My achievements, relationships, and the sheer wonder of existence provided all the meaning I sought. To me, the universe was a vast, intricate system governed by natural laws—a breathtaking interplay of forces and phenomena. Human existence, in all its complexity, was remarkable in its own right. This perspective offered clarity and a deep sense of direction, grounding me in the here and now while inspiring me to make the most of each fleeting moment.

So, why did I revisit my views on Creation and existence? It wasn't out of lack, longing, or a desire to fill an emotional or spiritual void. I wasn't searching for purpose—I already had that in abundance. Instead, my shift was born from a sincere and rational pursuit of truth. It was a journey guided by curiosity, science, evidence, and reflection—a quest to uncover what reality might reveal, regardless of where that exploration led.

Now, as someone who believes my life is part of a wondrous Creation, has my life changed? Unequivocally, yes. My perspective has expanded in ways I could never have anticipated. I now see the world as a breathtaking masterpiece, imbued with intentionality and design. Everything I once admired for its beauty or complexity now resonates with a profound sense of purpose.

The intricate harmony of Creation fills me with awe—from the delicate patterns on a blade of grass to the staggering complexity of the universe itself. What once seemed like random beauty now appears as a deliberate and unparalleled expression of creativity and intelligence. Each day feels infused with a renewed sense of wonder. The sunrise is no longer just a play of light and color; it's a testament to the precision and order governing existence. The laughter of my grandchildren isn't merely joyful noise—it's a poignant reminder of life's miracles. Each giggle carries the promise of love and hope passed down through generations of my family, affirming the resilience and beauty of life even in the face of challenges. The natural world, in its infinite complexity and interdependence, now feels like a living testament to something far greater than I had previously acknowledged. Where I once saw randomness, I now see patterns and meaning.

Where I once marveled at nature's beauty, I now find myself contemplating the Creation behind it. This shift has brought a deeper dimension to my existence. Every moment, every interaction, and every experience feels charged with significance. Life is no longer just something to enjoy; it's something to explore, celebrate, and honor.

That said, what does the recognition of the reality of Creation truly signify—and can we ever fully grasp its meaning? As I reflect on this, my mind often returns to a memorable scene from a film I deeply enjoyed years ago: the science fiction film *Contact* (1997), based on Carl Sagan's 1985 novel of the same name. The film follows Dr. Ellie Arroway, a passionate scientist working with SETI. In the film, Ellie is chosen to represent humanity and is transported in a spacecraft / time machine through a series of wormholes. During her journey, she has a profound encounter with an alien intelligence, which communicates with her in a deeply personal and transformative way. In the astonishing experience, Dr. Arroway expresses an awe that words cannot fully capture, declaring, "They should have sent a poet. It's so beautiful. So beautiful... I had no idea."

Her exclamation reflects the overwhelming beauty and majesty of the cosmos she has just witnessed. It underscores the idea that some experiences transcend the descriptive power of science and require the emotional depth and resonance of art or poetry to convey. Arroway's statement, brimming with reverence and humility, reveals a profound truth. Confronted by the ineffable vastness and elegance of the universe, she articulates a realization that transcends rational analysis. Certain mysteries defy the frameworks we often use to understand the world—mathematics, physics, or logic. To grasp such mysteries in their fullness, perhaps what is needed is not just a physicist or mathematician, but a poet. Her declaration reminds us that the universe is as much a source of wonder and inspiration as it is a subject for empirical inquiry.

As I emphasized in the introduction, this book has been written primarily for skeptics of a Creation, inviting them to take a fresh look at the evidence. But also, this book is not written to challenge personal religious faith but to provide an alternative view—one that respects belief while engaging skeptics with reason and evidence. On this note, I wish to address readers of religious faith. While I do not personally subscribe to any religious belief, I hold deep respect for those who do. Faith provides countless individuals with structure, hope, and meaning, playing a vital role in shaping human culture and shared values. My intent here is not to challenge or diminish anyone's beliefs but rather to offer an alternative perspective on Creation—one that seeks to transcend dogma while honoring the profound questions and mysteries that religions have endeavored to explore throughout the ages.

The universe, as revealed through science, is astonishing in its complexity, precision, and vastness. From the Big Bang onward, every step in the cosmic journey suggests an unfolding story imbued with intent and purpose. Skeptics often argue that apparent design can be explained by natural random processes. These explanations, while intriguing, fail to address the deeper question: why do these processes exist at all? The idea that pure chance could explain the fine-tuning of the universe isn't just unlikely—it's like believing in magic. But not the fun, storybook kind. It's a strange kind of "scientific magic" that says order, complexity, and life just happened—without reason, design, or direction. This way of thinking stems from a false divergence: the belief that accepting Creation necessitates accepting religion. But as I have demonstrated, this is simply not true. It is entirely possible to see the universe as a deliberate, intentional act without adhering to the tenets of organized religion.

Science excels at uncovering the mechanisms of the universe: the laws that govern particles, the forces that shape galaxies, and the processes that drive life's evolution. Yet, science often stops short of addressing the larger why. Why do these laws exist? Why is the universe comprehensible? Why does it inspire awe? This is where art and poetry enter the conversation. Art provides a language for the intangible—for emotions, beauty, and the sublime. When Dr. Arroway exclaims, "They should have sent a poet," she acknowledges that some aspects of existence are best understood not through equations but through metaphor and imagery. Einstein's words highlight the complementary nature of science and art. Where science dissects, art integrates; where science seeks clarity, art celebrates ambiguity.

Imagine holding a single piece of a jigsaw puzzle. You can examine its colors and shape, but without seeing the entire puzzle, it's impossible to grasp the full picture. A puzzle piece represents a fragment of understanding—what we know or perceive. Without the full puzzle, we cannot grasp the larger picture, much like how our limited knowledge about the universe prevents us from fully understanding its purpose or origins. This perspective urges us to approach the mysteries of the universe with humility. Every piece of knowledge we uncover, through scientific discovery, contributes to a larger, incomprehensible tapestry. Yet, the puzzle is far from complete, and perhaps it never will be. Instead of frustration, this realization can inspire wonder—an appreciation for the unknown as a vital part of existence itself.

One intriguing framework for interpreting the Creation event—especially when viewed through insights from the previous chapter—is the simulation hypothesis: the proposition that our reality may be a simulation orchestrated by an advanced intelligence.

While this sounds like a concept born purely from modern science fiction, a closer examination reveals a profound truth:

the simulation hypothesis is simply a contemporary dialect for an ancient reality. It is a way for our technological minds to conceptualize what a designed universe might actually look like. Whether the cosmos was brought into existence as a simulated digital construct or some other physical reality, the underlying architecture is identical. In both scenarios, the universe is an environment dictated by an external intellect, where the laws of nature are fixed parameters, and where matter is entirely subordinate to an underlying code. By exploring the ramifications of a simulation, we are fundamentally exploring the mechanics of *any* intentional creation.

By blending the mysteries of quantum mechanics with the idea of a deliberately designed simulation, this hypothesis offers a strikingly coherent way to interpret reality. The more I consider phenomena like measurement-dependent behavior, the universe's mathematical precision, and its seemingly information-based structure, the more this framework feels like the most rational explanation of the evidence we have. I know certainty is out of reach—but given the patterns we see, the simulation idea resonates with me as the most plausible interpretation.

If the universe is a simulation, then it is, by definition, a Creation. The Creator—an advanced civilization, or some other intelligence—designed this simulation with intent and purpose.

While it carries profound existential and philosophical implications, it does not necessarily have to be seen as unsettling. Regardless of whether reality is simulated, the pursuit of technological advancement, space exploration, and artificial intelligence remains deeply significant. The quest to understand the universe, improve human life, and expand knowledge retains its value, irrespective of the nature of reality itself. It even raises questions about the nature of death and what might follow. If our reality is a simulation, then death may not signify the end of our existence. Instead, it could represent a transition—our consciousness might be transferred to another simulated environment or preserved in some way by the simulation's Creator. This raises intriguing parallels with traditional religious concepts of an afterlife, suggesting that existence itself could be far more flexible and enduring than we typically imagine.

If our reality is a simulation, then everything we perceive—our physical world, our relationships, and even our belief systems—might ultimately be part of an elaborate construct. This would mean that religion, too, could have been intentionally programmed into the simulation by its Creator.

When viewed through this broader lens, ancient theological concepts cease to be mere dogmatic assertions and instead look like early human attempts to map out this designed reality. Throughout history, thinkers have used the language of religion to grapple with the boundaries, rules, and constraints of our existence. If we strip away the denominational imagery, many core tenets of faith begin to look remarkably like insights into an underlying, pre-programmed architecture.

The simulation hypothesis also provides a fresh lens through which to consider theological ideas. For example, Christianity's claim that salvation is a gift from God, rather than something achieved through human effort, aligns intriguingly with the notion of a simulated reality. If free will is itself an illusion—merely a programmed feature of the simulation—then the concept of relying on divine grace rather than personal merit takes on profound significance. In a universe where our choices may be preordained or heavily influenced by the constraints of a pre-designed system, trust in the Creator's wisdom and provision becomes not just a comfort but a necessity. This perspective challenges the traditional emphasis on individual agency, suggesting instead that our lives unfold within a meticulously crafted framework where every event, no matter how incomprehensible, serves a deliberate purpose.

The question of how we respond to life's hardships becomes deeply significant in this context. The parable of Job, found in the Bible, offers a timeless and thought-provoking exploration of this dilemma. Job, a man of exceptional righteousness and integrity, endures overwhelming suffering without apparent cause. In his anguish, he boldly questions God, seeking an explanation that aligns with his understanding of justice and fairness. Yet, rather than providing a direct answer, God responds with a poetic and awe-inspiring rebuke, unveiling the grandeur of creation and the unfathomable depth of divine wisdom. This response highlights the vast gulf between human understanding and the Creator's perspective, challenging Job—and us—to embrace humility and faith in the face of life's profound uncertainties. Both the parable of Job and the simulation hypothesis converge on this central truth: that peace is found not in definitive answers, but in trusting the wisdom and intention of the Creator.

During a business trip years ago, I had a conversation with a gentleman that profoundly impacted me. He shared the story of his daughters, both of whom had struggled with infertility for years. When one of them finally gave birth to a beautiful baby boy, the family's joy was immeasurable. But their happiness was tragically short-lived. At just two years old, the little boy was diagnosed with brain cancer and passed away. As the man recounted this heartbreaking loss, his voice trembled, and I could see the weight of the grief he carried. I fought back tears, moved by his courage and his faith. Despite their immense sorrow, the family chose to trust in what they saw as God's wisdom. They believed that, though they could not understand it, their suffering had a purpose beyond their grasp. Their unwavering faith reminded me of Job's story—a testament to the power of acceptance, even in the face of unimaginable pain.

There is a deep and thought-provoking irony in the ideas that emerge when we consider the simulation hypothesis, or even what we currently understand about the quantum world. Everything we experience—our universe, its history, and even our sense of self—might be a meticulously constructed illusion. If this were true, it raises profound questions: how could we ever be certain of the origins of anything, including the age or authenticity of the universe itself? We perceive the universe as ancient, approximately 13.8 billion years old, with its vast history etched into the stars. Yet, if this reality is a simulation, what we consider evidence could be nothing more than a sophisticated preloaded dataset. The universe might not have unfolded over billions of years but instead could have been instantiated with a fabricated history to give the illusion of depth.

Quantum mechanics lends credibility to this unsettling proposition. The mere act of measuring a quantum system changes its state. This phenomenon suggests a world that doesn't exist in a definite state until it is measured, as if reality itself requires an observation, to "collapse" possibilities into a single outcome. Similarly, quantum entanglement challenges the very fabric of space and time. Two particles can become so deeply linked that the state of one instantly influences the other, no matter the distance between them. In the context of a simulation, such phenomena could reflect shortcuts within the system—allowing for instantaneous communication between points that appear to be light-years apart. The malleability of the quantum world further reinforces the idea that our reality might be programmed. These quantum oddities suggest a universe that behaves less like a fixed, objective reality and more like an adaptive and responsive system—a hallmark of a sophisticated simulation.

If our universe is indeed a simulation, it forces us to reconsider fundamental concepts like time, causality, and the very nature of existence. Time might be an artificial construct designed to provide coherence to the simulation's narrative. Moreover, quantum mechanics challenges our understanding of cause and effect. The delayed-choice experiment shows that decisions made in the present can retroactively influence events in the past. In a simulated universe, such paradoxes might reflect the system's ability to retroactively rewrite its own history to maintain internal consistency. The implications are profound. If the universe is a simulation, our pursuit of scientific truth might be less about uncovering an ultimate reality and more about exploring the rules and boundaries of the construct. Quantum weirdness becomes less of an enigma and more of a feature—a glimpse into the computational underpinnings of a reality designed to challenge, inspire, and perhaps even deceive its inhabitants.

This perspective reframes existence itself, suggesting that the universe is a purposeful and intricate illusion. Whether we are uncovering the "code" of this simulation or simply testing its limits, quantum mechanics offers tantalizing evidence that reality is far more malleable than we ever imagined. Surprisingly, this idea is not as far-fetched as it might initially seem. The simulation hypothesis finds support in advancements in artificial intelligence and virtual reality. Our own technology is rapidly approaching the point where we can create lifelike simulations. If we are already experimenting with such constructs, it's conceivable that a far more advanced civilization could have done the same on an incomprehensibly larger scale. If the simulation was designed to give us the impression of truth, what is its purpose? Are we part of an experiment, a test, or a work of art?

This leads to profound existential questions about the nature of meaning and purpose. If everything we know is an illusion, does that diminish its value? Or does the act of experiencing reality, however constructed, give it intrinsic worth? Perhaps the illusion of reality is all we need to find meaning, even if the ultimate "truth" remains out of reach. Ultimately, the simulation hypothesis offers a humbling reminder of the limits of human understanding. Whether or not we live in a simulation, the exercise of questioning our reality encourages us to approach existence with curiosity, humility, and wonder.

Another intriguing perspective is the idea that the universe was designed by a Creator—not as an end in itself but as a platform to spark further creative ideas. In this view, the purpose of life is fundamentally about generating the new. Creativity is the engine of progress, innovation, and evolution. This concept reframes our existence as an active participant in a grand experiment aimed at unlocking untapped potential.

Creativity in humans is often constrained by the frameworks of knowledge and experience. What if the Creator understood this limitation and sought to circumvent it? By designing a universe populated by autonomous beings capable of independent thought, the Creator may have aimed to introduce an element of unpredictability. Life becomes a kind of creative engine—one capable of generating ideas, inventions, and paradigms that even the Creator might not have anticipated. Humanity represents a pinnacle of this process. Our achievements often spring from the collision of ideas, resulting in something greater than the sum of its parts. In this framework, the Creator provides the conditions for life to flourish and innovate, then observes the novel outcomes that emerge. This notion aligns with the idea that creation is a process of self-discovery. By allowing life to unfold freely, the Creator enables the birth of forms of beauty and intelligence that might never have arisen from deliberate design alone.

By embedding randomness and freedom into the fabric of existence, the Creator ensures that life has the flexibility to produce the unexpected. The universe, then, becomes a laboratory of ideas. This perspective carries profound implications: our capacity to imagine and innovate is a central aspect of the Creator's design. We are not merely products of a designed system but essential contributors to an ongoing, unfolding creative process. Our discoveries and our art are the very things the universe was designed to produce. It suggests a collaborative partnership where the Creator provides the framework, but it is life that fills it with meaning and originality. We are co-creators in an ongoing process of discovery.

Out of necessity, we cannot—and actually should not—know everything, it is better that keep wondering. If we possessed absolute knowledge, particularly about something like the afterlife, it could create a profound paradox. For instance, documented cases exist of individuals who took their own lives because they believed—not merely hoped, but knew—that a perfect afterlife awaited them. This paradox illustrates why a "better beyond" must remain a mystery; certainty could shift death from a feared inevitability to a sought-after release, undermining the importance of the life we are currently living.

Religious beliefs reflect humanity's attempts to grapple with these mysteries. Religion, with its rituals and moral frameworks, serves as a way for humans to connect with each other and with what some choose to believe as the divine. These stories often contain profound truths, even if they are not always literal, representing our longing to connect with something greater than ourselves.

As Dr. Arroway suggests, it may ultimately be the poet, not the physicist, who helps us grasp the universe's meaning. While science provides the facts, poetry gives voice to the feelings those facts evoke. The poet captures the ineffable—the awe of a starry sky, the wonder of a newborn's breath. The poet's gift is in connecting the known and the unknown, reminding us that some truths cannot be measured but only felt. For all our equations, the universe remains a source of wonder. To embrace that wonder is to accept that we may never fully understand—and that is okay. The act of seeking and marveling is itself a profound expression of Creation.

As we conclude this journey, I invite you to reflect on the idea that a Creator exists and that we are created beings. The purpose and reason behind this Creation remain open to interpretation. The journey of understanding is not about arriving at definitive answers; rather, it's about embracing the beauty and complexity of the questions themselves. Our knowledge is akin to a puzzle piece, part of a larger tapestry that may forever elude our full comprehension. Yet, with curiosity and humility, we can continue to marvel at the mystery. Creation is a reality supported by evidence—a testament to the potential of existence. And within that mystery, we discover the potential for meaning, connection, and wonder. As Albert Einstein so eloquently observed, "We do not observe nature as it actually exists but nature exposed to our methods of perception." May these words inspire us to approach the mystery of Creation with reverence, curiosity, and a willingness to explore.

ACKNOWLEDGEMENTS

First and foremost, I want to express my deep gratitude to the field of science and the countless researchers whose dedication has unlocked the mysteries of our universe. Their tireless pursuit of truth and unyielding curiosity continue to illuminate the path toward understanding our place within the cosmos. This book is, in many ways, a celebration of their work; without the foundation of their discoveries, this project could not exist. I hope these pages do justice to their monumental contributions to humanity.

Beyond the scientific community, I am grateful to the many individuals who contributed to this book, often in ways they may never realize. Many of those who influenced this work may not even know of its existence, yet their perspectives are woven into its very fabric.

I owe much of my outlook to my father, Benjamin Windsor. Though he passed away some years ago, his impact remains profound. He was a complex, fascinating man who thrived on open dialogue. In our home, no topic was off-limits and no "sacred cows" went unchallenged. That culture of openness fueled my intellectual curiosity and taught me the invaluable lesson of respecting differing views while engaging deeply in philosophical debate.

I am also profoundly grateful to the thousands of people I've encountered during forty-five years of business travel—specifically, the farmers. By my best estimate, I have spoken with nearly 60,000 farmers over the decades. These encounters were extraordinary. Those who work the land, especially on the vast prairies, lead lives defined by resilience and a unique brand of introspection.

Time and again, I found myself in candid conversations with these men and women about religion, philosophy, and the complexities of life. Whether they were people of deep faith or secular thinkers, their honesty provided a wellspring of insight. The seeds for much of what is explored in this book were planted in those fields and farmhouse kitchens.

To my extended family, thank you for your unwavering support. Your patience and encouragement provided the solid ground I needed to dedicate myself to the years of research and writing required for this project.

Finally, my thanks go to Jean Moses for her invaluable advice and editing assistance. To all of you—scientists, family, friends, and the strangers who shared a moment of honest conversation—

my gratitude is immeasurable. This book exists because of your insight and inspiration.

Glossary of Terms

Abiogenesis:
The hypothesis that life arose from non-living chemical processes. Abiogenesis examines how early Earth's environment could have enabled simple molecules, like amino acids, to organize into complex, self-replicating systems that eventually evolved into living organisms.

Abiotic Synthesis:
The natural formation of organic compounds, such as amino acids and nucleotides, in environments devoid of life. This process is a foundational step in theories of life's origin and has been simulated in experiments like the Miller-Urey experiment.

Albert Einstein:
A physicist whose theories of Special and General Relativity transformed our understanding of space, time, and gravity. Einstein's work introduced groundbreaking ideas like spacetime curvature, gravitational waves, and energy-mass equivalence. His philosophical debates on quantum mechanics reflect his search for universal order.

Amino Acids:
Organic molecules that are the building blocks of proteins. Amino acids play essential roles in cellular processes and structure. Their presence in meteorites and laboratory experiments suggests they could form naturally under the right conditions.

Anthropic Coincidences:
Observable phenomena or constants in the universe that seem

finely tuned to support life, such as the strength of gravity or the properties of water. These coincidences often fuel debates about the intentionality of the universe's design.

Anthropic Principle:
The idea that the universe's physical laws and constants appear to be finely tuned to allow for the existence of conscious observers. This principle raises questions about whether the universe's conditions are intentional or coincidental.

Apoptosis, often called programmed cell death, is a highly regulated and controlled process that allows cells to die in a manner that benefits the organism. This mechanism is essential for maintaining cellular homeostasis, eliminating damaged or unnecessary cells, and ensuring proper development and functioning of multicellular organisms.

Astrobiology:
The interdisciplinary study of the potential for life elsewhere in the universe. Astrobiology combines elements of astronomy, biology, and chemistry to explore conditions for habitability and search for extraterrestrial life.

Atomic Structure:
The arrangement of protons, neutrons, and electrons in atoms, which form the building blocks of matter. Quantum mechanics governs atomic behavior, leading to chemical reactions that sustain life.

Ayn Rand:
A philosopher and novelist known for her advocacy of Objectivism, which promotes reason, individualism, and atheism. Rand's work critiques religion as irrational and emphasizes human potential through rational self-interest.

Base Pair:
A unit of DNA's double-helix structure, formed by the pairing of nucleotides adenine with thymine and cytosine with guanine.

These pairs encode genetic information crucial for biological replication and function.

Bell's Theorem:
A principle in quantum mechanics demonstrating that no local hidden variable theory can fully explain the behavior of entangled particles. Bell's theorem supports the inherent interconnectedness of quantum systems.

Big Bang Nucleosynthesis:
The formation of light elements such as hydrogen, helium, and lithium in the early moments of the universe. These elements provided the foundation for the development of stars and galaxies.

Big Bang Theory:
The prevailing cosmological model for the universe's origin, describing its expansion from an infinitely dense and hot singularity approximately 13.8 billion years ago. The Big Bang explains cosmic background radiation, the distribution of galaxies, and the abundance of light elements.

Biochemical Pathways:
Networks of chemical reactions within cells that sustain life by converting energy and synthesizing molecules. Examples include photosynthesis and cellular respiration, which highlight the efficiency and complexity of life's systems.

Biogenesis Law:
The principle that life originates only from pre-existing life, contrasting with theories of abiogenesis. This law underscores the continuity of living systems while highlighting the mystery of life's ultimate origins.

Block Universe:
A theoretical concept in which all moments in time—past, present, and future—exist simultaneously within a four-dimensional spacetime continuum. This idea, rooted in Einstein's relativity, challenges our perception of time as linear.

Carbon-Based Life:
Life forms that rely on carbon's unique ability to form stable, complex molecules like DNA, proteins, and carbohydrates. Carbon's versatility makes it foundational to all known biological systems.

Causality:
The principle that every effect must have a cause. This foundational concept underpins scientific inquiry and fuels philosophical debates about the origins of the universe and its first cause.

Cellular Complexity:
The intricate structures and processes within cells, from organelles like the mitochondria to molecular machines like ribosomes. This complexity showcases life's sophistication and raises questions about its origin.

Chemical Evolution:
The process by which simple chemical compounds gradually formed more complex molecules, potentially leading to life. This transition bridges abiotic chemistry and biological systems.

Christopher Hitchens:
A journalist, author, and outspoken atheist known for his critique of religion in works like *God Is Not Great*. Hitchens argued for secularism and humanism, often debating the role of faith in modern society.

Circumstellar Habitable Zone:
The region around a star where conditions are "just right" for liquid water to exist, often referred to as the "Goldilocks zone." This concept is critical in the search for habitable exoplanets.

Cosmic Evolution:
The study of how the universe changes over time, from the Big Bang to the formation of galaxies, stars, planets, and ultimately life. Cosmic evolution integrates physics, astronomy, and biology to explore the universe's development.

Cosmic Horizon Problem:
A cosmological puzzle regarding why the Cosmic Microwave Background Radiation is so uniform across vast regions of the universe, despite them being causally disconnected.

Cosmic Inflation:
A theory describing a rapid expansion of the universe in its first fractions of a second. Inflation resolves cosmological puzzles like the horizon and flatness problems.

Cosmic Microwave Background Radiation (CMB):
The faint glow of radiation left over from the Big Bang, providing a snapshot of the universe's early state and serving as critical evidence for its origin.

Cosmological Constant (Λ):
A term in Einstein's equations representing the energy density of space. It explains the universe's accelerated expansion and exemplifies fine-tuning in cosmology.

Cosmology:
The scientific study of the universe's origin, evolution, structure, and ultimate fate. Cosmology combines theoretical physics with observational data to address fundamental questions about existence.

Creation:
The act of bringing the universe into existence. In the context of this book, Creation is examined through scientific theories such as the Big Bang and philosophical arguments about intentionality, purpose, and the existence of a Creation.

Creation:
An intelligent force or entity posited to have designed and initiated the universe. The concept of a Creation is central to debates on the fine-tuning of the universe and the origins of life.

Dark Energy:
A mysterious form of energy that drives the accelerated expansion of the universe. Dark energy constitutes approximately 68% of the universe's total energy and remains one of the most significant mysteries in modern cosmology.

Dark Matter:
A form of matter that does not emit or interact with electromagnetic radiation, making it invisible to telescopes. Inferred from its gravitational effects, dark matter is believed to make up about 27% of the universe's total mass-energy content.

David Attenborough:
A naturalist and broadcaster whose documentaries, such as *Planet Earth*, explore the diversity and complexity of life. His work emphasizes the interconnectedness of ecosystems and often sparks reflections on evolution and the origins of life.

Deoxyribonucleic Acid (DNA):
The molecule that carries genetic information in all known living organisms. DNA's double-helix structure enables it to store, replicate, and transmit genetic instructions, making it essential for life's continuity.

Determinism:
The philosophical concept that all events are determined by preceding causes, leaving no room for randomness. Quantum mechanics introduces probabilistic behavior that challenges classical determinism.

Double-Helix Structure:
The spiral-shaped structure of DNA, discovered by Watson and Crick in 1953. This elegant configuration enables the storage and accurate replication of genetic information.

Double-Slit Experiment:
A foundational quantum mechanics experiment demonstrating wave-particle duality. Particles, like electrons or photons, exhibit

wave-like interference patterns unless observed, challenging classical ideas about reality.

Einstein-Rosen Bridge:
A theoretical concept in General Relativity describing a "shortcut" through spacetime, commonly known as a wormhole. While speculative, it has inspired discussions about time travel and the nature of spacetime.

Electromagnetic Force:

One of the four fundamental forces of nature, responsible for the interactions between charged particles. It governs phenomena such as light, electricity, and magnetism, mediated by photons.

Emergent Phenomena:
Complex behaviors or structures that arise from the interactions of simpler components. Examples include the self-organization of galaxies and the emergence of consciousness from neural activity.

Entropy:
A measure of disorder in a system. The Second Law of Thermodynamics states that entropy tends to increase over time, influencing the universe's evolution and the concept of time's arrow.

Epigenetics:
The study of heritable changes in gene expression that do not involve alterations to the DNA sequence. Environmental factors often influence epigenetic modifications, demonstrating the dynamic interaction between genetics and external conditions.

Error-Correcting Mechanisms:
Biological processes that detect and repair errors in DNA during replication. These mechanisms maintain genetic integrity and highlight the precision of molecular biology.

Exoplanets:
Planets that orbit stars outside our solar system. The discovery of

thousands of exoplanets has expanded the search for potentially habitable worlds and deepened our understanding of planetary systems.

Extraterrestrial Intelligence:
The theoretical existence of intelligent life beyond Earth. The search for extraterrestrial intelligence (SETI) investigates signals or other evidence of advanced civilizations in the universe.

Fine-Tuned Constants:
The precise values of physical constants, such as the speed of light and Planck's constant, that allow for a stable and life-permitting universe. These constants are often central to discussions about intentionality and design.

Fine-Tuning:
The observation that the universe's physical laws and constants are delicately balanced to allow for the existence of life. Fine-tuning is often cited as evidence for a Creation or multiverse theories.

Fibonacci Sequence:
A mathematical sequence where each number is the sum of the two preceding numbers. This sequence appears in natural patterns such as the arrangement of leaves, spirals in galaxies, and the structure of pinecones, reflecting mathematical elegance in nature.

Francis Collins:
A geneticist who directed the Human Genome Project and advocates for the compatibility of faith and science. Collins' book *The Language of God* explores his journey from atheism to belief in a Creation.

Fred Hoyle:
An astrophysicist who criticized the Big Bang theory but famously argued that the fine-tuning of the universe suggests the involvement of a "super intellect."

General Theory of Relativity:
Einstein's groundbreaking theory describing gravity as the warping of spacetime by mass and energy. General Relativity has been confirmed through observations of phenomena like gravitational waves and black holes.

Gluon: The particle that acts as the exchange particle for the strong nuclear force, binding quarks together.

Gravitational Constant (G):
A fundamental constant representing the strength of gravitational attraction between masses. Its precise value is crucial for the stability of planetary systems and the formation of stars and galaxies.

Gravitational Waves:
Ripples in spacetime caused by massive accelerating objects, such as merging black holes. Their detection in 2015 confirmed a prediction of Einstein's General Relativity.

Hafele-Keating Experiment:
An experiment conducted in 1971 that used atomic clocks aboard airplanes to confirm time dilation, a key prediction of Einstein's theory of relativity.

Heat Death of the Universe:
A theoretical scenario in which the universe reaches maximum entropy, with all energy evenly distributed. This would result in a state where no usable energy remains to sustain processes or structures.

Helium Abundance:
The proportion of helium in the universe, formed during Big Bang nucleosynthesis. Its observed levels provide key evidence for the Big Bang model.

Higgs Boson:
A particle associated with the Higgs field, which gives mass to

other particles. Its discovery in 2012 at CERN validated a crucial aspect of the Standard Model of particle physics.

James Tour:
A chemist and nanotechnologist known for his critiques of abiogenesis theories. Tour emphasizes the complexity of molecular systems as evidence against naturalistic explanations for the origin of life.

Jean-Paul Sartre:
A philosopher associated with existentialism, emphasizing individual freedom and the absence of inherent meaning in the universe. Sartre rejected religious frameworks and encouraged humans to create their own purpose.

John Lennox:
A mathematician and Christian apologist who argues for the compatibility of science and faith. Lennox often debates prominent atheists, presenting philosophical arguments for the existence of a Creation.

Kalam Cosmological Argument:
A philosophical argument asserting that everything that begins to exist has a cause. This argument is central to William Lane Craig's defense of theism and posits a Creation as the universe's ultimate cause.

Lawrence Krauss:
A physicist and cosmologist who argues that quantum mechanics and cosmology can explain the universe's origin without invoking a Creation. His book *A Universe from Nothing* explores these ideas in depth.

Lepton: A fundamental particle that does not experience the strong nuclear force, such as electrons, muons, and neutrinos.

LIGO (Laser Interferometer Gravitational-Wave Observatory):
A facility that detects gravitational waves, providing insights into black holes, neutron stars, and the fabric of spacetime.

Light-Year:
The distance light travels in one year, approximately 5.88 trillion miles. Light-years are used to measure astronomical distances, providing a sense of scale for the universe's vastness.

Madalyn Murray O'Hair:
An atheist activist who played a key role in removing mandatory prayer and Bible readings from U.S. public schools. O'Hair was a polarizing figure and founder of American Atheists, advocating for the separation of church and state.

Multiverse Hypothesis:
The idea that our universe is one of many, each with different physical laws and constants. This hypothesis is proposed to explain the fine-tuning of the universe without invoking a Creation.

Natural Laws:
The fundamental principles governing the behavior of matter and energy, such as gravity and electromagnetism. These laws are consistent throughout the universe, forming the foundation of scientific understanding.

Natural Selection:
The process by which traits that enhance survival and reproduction become more common in a population over generations. Natural selection drives evolution and explains the adaptation of species to their environments.

Neutrinos:
Nearly massless subatomic particles that rarely interact with matter. Neutrinos are produced in nuclear reactions, such as those in stars, and provide valuable insights into the universe's inner workings.

Nucleosynthesis: the process by which atomic nuclei are created, typically within stars, during various stages of their life cycles. It is a fundamental concept in astrophysics that explains the origin of the elements in the universe.

Observable Horizon:
The maximum distance from which light has had time to reach Earth since the universe began. This limit defines the observable universe and its boundaries.

Observable Universe:
The portion of the universe visible from Earth, limited by the speed of light and the age of the universe. It spans approximately 93 billion light-years in diameter.

Pale Blue Dot:
A term coined by Carl Sagan to describe Earth as seen from deep space, emphasizing its fragility and the interconnectedness of all life.

Photon: The force carrier of the electromagnetic force and the quantum of light.

Planck Constant (h):
A fundamental constant in quantum mechanics that relates a particle's energy to its frequency. It defines the scale at which quantum effects dominate.

Planck Epoch:
The earliest phase of the universe, lasting up to 10−4310^{-43}10−43 seconds after the Big Bang. During this period, quantum gravitational effects were significant, and current physics theories cannot fully describe its conditions.

Planck Length:
The smallest measurable length, approximately 1.6×10−351.6 \times 10^{-35}1.6×10−35 meters. At scales smaller than the Planck length, classical physics breaks down, and quantum effects dominate.

Planck Time:
The smallest measurable unit of time, approximately 5.39×10−445.39 \times 10^{-44}5.39×10−44 seconds. It

represents the earliest period of the universe immediately after the Big Bang.

Primordial Density Fluctuations:
Tiny variations in the density of matter in the early universe, which served as the seeds for the formation of galaxies and large-scale cosmic structures.

Protein Synthesis:
The biological process by which DNA instructions are translated into functional proteins. This process, involving transcription and translation, is central to cellular function and life's complexity.

Quantum Chromodynamics (QCD):
A theory describing the strong nuclear force, which binds quarks together in protons and neutrons. QCD is a key component of the Standard Model of particle physics.

Quantum Coherence:
A property of quantum systems where particles exhibit correlated behavior across distances. Coherence is essential to phenomena like entanglement and quantum computing.

Quantum Computing:
A revolutionary field of computing that leverages quantum states, such as superposition and entanglement, to perform calculations far beyond the capabilities of classical systems.

Quantum Decoherence:
The process by which quantum systems lose their quantum properties due to interaction with their environment, resulting in classical behavior.

Quantum Entanglement:
A phenomenon where particles remain interconnected, so the state of one instantly influences the state of another, regardless of distance. Entanglement challenges classical physics and underscores the interconnectedness of quantum systems.

Quantum Foam:
A theoretical description of spacetime at the smallest scales, where quantum fluctuations cause it to appear chaotic and dynamic.

Quantum Gravity:
A theoretical framework that seeks to unify quantum mechanics and general relativity, addressing phenomena such as black holes and the universe's origin.

Quantum Mechanics:
The branch of physics studying the behavior of particles at atomic and subatomic scales. Quantum mechanics reveals counterintuitive principles like superposition, wave-particle duality, and entanglement.

Quantum Nonlocality:
A feature of quantum mechanics where particles influence each other instantly over vast distances, defying classical notions of locality.

Quantum Superposition:
The principle that particles exist in multiple states simultaneously until measured. Superposition underlies many phenomena in quantum mechanics, including the famous Schrödinger's Cat thought experiment.

Quantum Tunneling:
A quantum effect where particles pass through barriers that would be impassable under classical physics. This phenomenon is crucial for processes like nuclear fusion in stars.

Quark: A fundamental particle and a building block of matter, combining to form protons and neutrons. Quarks come in six "flavors": up, down, charm, strange, top, and bottom.

Radiometric Dating:
A technique for determining the age of materials by measuring the

decay of radioactive isotopes. Radiometric dating provides timelines for Earth's history and the evolution of life.

Relativity:
Einstein's theory describing the relationships between space, time, and gravity. Relativity explains phenomena such as time dilation, black holes, and the expansion of the universe.

Richard Dawkins:
An evolutionary biologist and outspoken atheist known for his critiques of religion and advocacy for science-based explanations of life's complexity. His book *The God Delusion* has been influential in secular and scientific communities.

Schrödinger's Cat:
A thought experiment illustrating quantum superposition, where a cat in a box can be simultaneously alive and dead until observed. It highlights the paradoxical nature of quantum mechanics.

Second Law of Thermodynamics:
A fundamental principle stating that entropy, or disorder, in an isolated system always increases over time. This law governs the direction of natural processes and energy flow.

Self-Replicating Molecules:
Molecules capable of copying themselves, such as RNA. These molecules are considered essential to the origin of life and the transition from chemistry to biology.

Singularities:
Points of infinite density and curvature, such as those at the center of black holes or the origin of the universe. Singularities challenge current physical theories.

Solar System Formation:
The process by which the Sun and its planets formed from a collapsing molecular cloud approximately 4.6 billion years ago.

Space-Time Continuum:
The four-dimensional framework of space and time described by Einstein's relativity, within which all physical phenomena occur.

Standard Model of Particle Physics:
A theoretical framework describing the fundamental particles and forces, excluding gravity, that govern the universe.

String Theory:
A theoretical framework suggesting that the fundamental building blocks of the universe are one-dimensional strings vibrating at specific frequencies. String theory aims to unify quantum mechanics and general relativity.

Supermassive Black Holes:
Enormous black holes found at the centers of most galaxies, including our Milky Way. These objects play a crucial role in galaxy formation and evolution.

Time Dilation:
A relativistic effect where time slows down for objects moving at high velocities or in strong gravitational fields. Time dilation has been confirmed experimentally and is essential to modern physics.

Time's Arrow:
The concept describing the unidirectional flow of time, linked to increasing entropy as outlined by the Second Law of Thermodynamics.

Universal Constants:
Fundamental physical values, such as the speed of light and Planck's constant, that define the behavior of the universe. Their precise calibration is central to the fine-tuning argument.

Vacuum Energy:
The energy inherent in empty space, contributing to the universe's accelerated expansion. Vacuum energy is a key component of quantum field theories.

Wavefunction:
A mathematical representation of a quantum system, describing the probabilities of its possible states. Measurement causes the wavefunction to collapse into a definite state.

Wave-Particle Duality:
The principle that particles like photons and electrons exhibit both wave-like and particle-like properties, depending on how they are observed.

Werner Heisenberg:
A physicist who developed the Uncertainty Principle, foundational to quantum mechanics. His work challenges classical notions of determinism and measurement.

White Dwarfs:
Dense remnants of stars that have exhausted their nuclear fuel, marking the final stages of stellar evolution for many stars.

William Lane Craig:
A philosopher and theologian known for the Kalam Cosmological Argument, which asserts that the universe's beginning necessitates a cause outside itself. Craig's work emphasizes philosophical and logical evidence for a Creation.

Zero-Point Energy:
The lowest possible energy state of a quantum system, even in the absence of matter or heat. This phenomenon has implications for quantum physics and cosmology.

Zygote:
The single cell formed by the fusion of sperm and egg, marking the beginning of a new organism's development. The zygote contains the complete genetic blueprint for life.

Made in United States
Troutdale, OR
11/04/2026

56942088R00090